THE INTERFACIAL TRANSITION ZONE IN CEMENTITIOUS COMPOSITES

Also available from E & FN Spon

Autogenous Shrinkage of Concrete
Edited by E. Tazawa

Concrete Materials
M. Levitt

Corrosion of Steel in Concrete
J.P. Broomfield

High Performance Concrete
P-C. Aitcin

Interfacial Transition Zone in Concrete
Edited by J.C. Maso

Interfaces in Cementitious Composites
Edited by J.C. Maso

Mechanisms of Chemical Degradation of Cement-based Systems
Edited by K.L. Scrivener and J.F. Young

Optimization Methods for Material Design of Cement-based Composites
Edited by A.M. Brandt

Penetration and Permeability of Concrete
Edited by H.W. Reinhardt

Polymers in Concrete
Edited by Y. Ohama

Prediction of Concrete Durability
Edited by J. Glanville and A. Neville

Prevention of Thermal Cracking in Concrete at Early Ages
Edited by R. Springenschmid

For information on these and any other books or journals published by E & FN Spon, please contact: The Marketing Department, E & FN Spon, 11 New Fetter Lane, London EC4P 4EE.

RILEM

Second International Conference on

THE INTERFACIAL TRANSITION ZONE IN CEMENTITIOUS COMPOSITES

Haifa, Israel

March 8–12, 1998

EDITED BY

A. Katz, A. Bentur
National Building Research Institute
Department of Civil Engeneering
Technion-Israel Institute of Technology
Haifa 32000, Israel

M. Alexander
Department of Civil Engineering
University of Cape Town
7700 Rondebosch
South Africa

G. Arliguie
Laboratoire Matèriaux et Durabilité des Constructions
UPS, Toulouse, France

CRC Press
Taylor & Francis Group
Boca Raton London New York

CRC Press is an imprint of the
Taylor & Francis Group, an **informa** business
A SPON PRESS BOOK

CRC Press
Taylor & Francis Group
6000 Broken Sound Parkway NW, Suite 300
Boca Raton, FL 33487-2742

First issued in paperback 2019

© 1998 RILEM
CRC Press is an imprint of Taylor & Francis Group, an Informa business

ISBN-13: 978-0-419-24310-6 (hbk)
ISBN-13: 978-0-367-86594-8 (pbk)

Publisher's Note
This book has been prepared from camera-ready copy provided by the author.

British Library Cataloguing in Publication Data
A catalogue record for this book is available
from the British Library

Library of Congress Cataloging in Publication Data
A catalogue record for this book has been requested

Visit the Taylor & Francis Web site at
http://www.taylorandfrancis.com

and the CRC Press Web site at
http://www.crcpress.com

Contents

Preface   xiii

Part One: Keynote Lecture

1   **The Interfacial Transition Zone: Reality or Myth?**   3
S. Diamond and J. Huang

Part Two: Modeling, Characterization and Quantification of ITZ Structure and Properties

2   **Concrete: A Multi-scale Interactive Composite**   43
E.J. Garboczi and D.P. Bentz

3   **Fracture of the Bond Between Aggregate and Matrix: An Experimental and Numerical Study**   51
A. Vervuurt and J.G.M. van Mier

4   **Introduction of Syneresis in Cement Paste**   59
M.R. De Rooij, J.M.J.M. Bijen and G. Frens

5   **Two-dimensional Concrete Models Using Metal Aggregates**   67
M. Suarjana, M.S. Besari and R. Abipramono

6   **Interfacial Transition and Destruction in Hardening of Water–Silicate Dispersion Systems**   75
D.I. Shtakelberg and S.V. Boiko

Part Three: Role of ITZ in Controlling Diffusion and Permeability

7   **Gas Permeability of Mortars in Relation with the Microstructure of Interfacial Transition Zone (ITZ)**   85
M. Carcasses, J.Y. Petit and J.P. Ollivier

8   **Modeling the Modification of Chloride Penetration in Relation to the ITZ Damage**   93
R. François, G. Arliguie and A. Konin

9 Influence of the Interfacial Transition Zone on the
   Resistance of Mortar to Calcium Leaching 103
   A. Delagrave, J. Marchand and M. Pigeon

10 Evaluation of the Interfacial Transition Zone in
   Concrete Affected by Alkali–Silica Reaction 114
   P. Rivard, G. Ballivy and B. Fournier

Part Four: Role of ITZ in Controlling Durability of Concrete

11 Influence of the Interfacial Transition Zone on the
   Frost Resistance of Calcareous Concretes 125
   F. Renaud-Casbonne, M. Bellanger and F. Homand

12 Granitic Rocks Attack by Alkali–Silica Reaction and
   Formation in the Cement Paste Interface 133
   E. Menéndez

13 Sulfate Attack, Interfaces and Concrete Deterioration 141
   J.P. Skalny, S. Diamond and R.J. Lee

14 Interfacial Effects in Glass Fiber Reinforced
   Cementitious Materials 152
   K. Kovler, A. Bentur and I. Odler

**Part Five: ITZ Around Reinforcing Bars and Its Role in Steel
Corrosion**

15 Action of Chloride Ions on the Reactions in the
   Corroded Steel–Cement Paste Interfacial
   Transition Zone 163
   J.L. Gallias and R. Cabrillac

16 Microstructure of the Interfacial Transition Zone
   Around Corroded Reinforcement 171
   J.L. Gallias

17 Effects of Interfacial Transition Zone Around
   Reinforcement on its Corrosion in Mortars With and
   Without Reactive Aggregate 179
   M. Kawamura, D. Singhal and Y. Tsuji

18 **Effects of the Reinforcement Corrosion Induced by Chlorides on the Interface Steel–Cement Paste in Mortars** 187
  E. Menéndez, M.A. Sanjuán and C. Andrade

19 **Comparison of ITZ Characteristics Around Galvanized and Ordinary Steel Rebars** 196
  F. Belaid, G. Arliguie and R. François

Part Six: ITZ Structure and Properties in Portland Cement Systems

20 **Appetency and Adhesion: Analysis of the Kinetics of Contact Between Concrete and Repairing Mortars** 207
  L. Courard and A. Darimont

21 **Restrained Shrinkage Cracking in Bonded Fiber Reinforced Shotcrete** 216
  N. Banthia and K. Campbell

22 **Features of the Interfacial Transition Zone and Its Role in Secondary Mineralization** 224
  D. Bonen

23 **Assessment of Cementitious Composites by In-Situ SEM Bending Test** 234
  J.M. Bartos and P. Trtik

Part Seven: ITZ Structure and Properties in Special Concrets

24 **Microhardness and Mechanical Behavior of the Expanded Shale Concrete** 243
  S.-W. Chen and U. Schneider

25 **Microcements as Building Blocks in the Microstructure of High Performance Concretes** 251
  D. Israel and W. Perbix

26 **Bond Property Studies on Ceramic Tile Finishes** 259
  Z.J. Li, W. Yao, M. Qi, S. Lee, X.S. Li and C.H. Lee

Part Eight: Effect of Chemical and Physical Properties of Aggregates on ITZ

27 **Interaction Between Fly-ash Aggregate With Cement Paste Matrix** 267
R. Wasserman and A. Bentur

28 **Properties of Interfacial Transition Zone in Gypsum-free Cement–Rock Composites** 276
Z. Bažantová and S. Modrý

29 **Influence of Initial Microstructuring in the ITZ on Early Hydration-strength Binding** 283
C. Legrand and E. Wirquin

30 **The Influence of Time-dependent Changes in ITZ on Stiffness of Concretes Made with Two Aggregate Types** 292
M.G. Alexander and K. Scrivener

31 **Interfacial Fracture Between Concrete and Rock Under Impact Loading** 301
S. Mindess and K.-A. Rieder

Part Nine: Effect of Additives and Admixtures on ITZ Structure and Properties of Concrete

32 **Characters of Interfacial Transition Zone in Cement Paste with Admixtures** 311
K. Kobayashi, A. Hattori and T. Miyagawa

33 **A Look at the Inner-surface of Entrapped Air Bubbles in Polymer–Cement–Concrete (PCC)** 319
M. Puterman

34 **The Role of Silica Fume in Mortars with Two Types of Aggregates** 324
A. Goldman and M.D. Cohen

Part Ten: Closing Conclusions

35   **ITZ Structure and Its Influence on Engineering and Transport Properties: Concluding Remarks to the Conference**   335
     A. Bentur

     Author Index   339

     Key Word Index   341

Sponsored by:

RILEM - International Union for Testing and Research
Laboratories for Materials and Construction

Technion- Israel Institute of Technology

Ministry of Science

Israel Readymixed Concrete Association

Thermokir Industries

Nesher – Israel Cement Enterprises Ltd.

Ashkalit – Ashtrom Industries

Conference Organization:
Chairman: A. Bentur
Secretary: A. Katz
          National Building Research Institute, Department of Civil Engeneering,
          Technion-Israel Institute of Technology, Haifa 32000, Israel

International Scientific Committee
M. Alexander (South Africa)
G. Arliguie (France)
G. Ballivy (Canada)
N. Banthia (Canada)
A. Bentur (Israel)
D. Bentz (U.S.A.)
J.P. Bournazel (France)
Y. Houst(Switzerland)
A. Katz (Israel)
B. Lagerblad (Sweden)
S. Mindess (Canada)
J.P. Ollivier (France)
K. Scrivener (France)
J. Van-Mier (Netherland)

Local Organizing Committee
M. Ben-Basat
A. Bentur
A. Goldman
A. Katz
K. Kovler

The conference was supported by:
RILEM –   International Union of Testing and Research Laboratories for
              Materials and Structure
Technion – Israel Institute of Technology
Israeli Ministry of Science
Israel Readymixed Concrete Association
Thermokir Industries
Nesher - Israel Cement Enterprises Ltd.
Ashkalit – Ashtrom Industries

PREFACE

Traditionally, the properties of the main cementitious material, concrete, are considered to be a function of the characteristics of the bulk cement paste matrix. This assumption formed the basis for the working hypothesis which enable adequate modeling of the behavior of traditional concrete mixes. In the last two decades a vast body of research has developed showing that the paste structure in concrete is not necessarily identical to that of the bulk paste, due to the formation of an interfacial transition zone (ITZ) around inclusions in cement paste (e.g. aggregates, fibres) where the microstructure of the paste is different than that of the bulk. The dependence of the ITZ structure on various parameters such as composition of the binder and the inclusion was studied and several methods have been developed to characterize the composition and structure of the ITZ and its changes as a function of the distance from the inclusion surface. The state of the art of the structure and characterization of the ITZ was the topic of the RILEM Technical Committee 108-ICC, Interface in Cementitious Composites. It published two reports, one the proceedings of a conference and the other a state of the art report. This committee completed its work in 1992.

A key question that needed to be resolved was to what extent the existence of ITZ had any practical influences on the engineering properties of cementitious materials, or was it just a peculiarity of academic interest. The question was brought up in view of the fact that for traditional concrete the properties could be accounted for by the bulk properties of the cementitious matrix. To resolve this issue two committees were set by RILEM, one dealing with engineering properties (TC-ETC: Engineering of the Interfacial Transition Zone in Cementitious Composites) and the other with transport properties which are related to durability (TC-TPZ: Transport Properties of the Interfacial Transition Zone). The two committees are coordinating their work, and this Symposium is jointly sponsored by both.

In discussing the practical influence of ITZ, one has to take into consideration that in the last two decades major strides have been taken to develop advanced high quality cementitious materials such as high strength concretes and fiber reinforced cements. These advanced materials are essentially composites where the inclusions in the cementitious matrix (aggregates and fibres) are not just inert fillers but are becoming active components in controlling the overall properties of the material. Thus, modern cementitious materials are essentially composites, and therefore one would expect that the special structure of the ITZ would play a much more important role than in the traditional concrete, where the aggregates were to a large extent diluting fillers. Therefore, our better understanding of the special nature of the ITZ on the one hand, and the development of advanced cementitious materials which are essentially composites on the other hand, are combining into a situation where the examination of the role of ITZ in controlling the properties of cementitious composites is advancing from an academic-theoretical issue to a practical one.

Thus, the main object of this symposium is to resolve the influence of ITZ on engineering and durability characteristics of cementitious composites, to identify what systems and which of the properties are affected by it, and to quantify these effects in order to prepare the base for engineering design tools.

Several topics are addressed in the Proceedings of this conference: quantification and characterization of the structure and properties of the ITZ itself, modeling these characteristics and their influence on the bulk behavior of various types of cementitious composites, durability characteristics and their dependency on ITZ in particular with respect to transport characteristics (diffusion and permeability), and the role of ITZ in specific systems: fiber reinforced cements, concretes with special aggregates, concretes with special admixtures, repair and bonding systems, and the influence of ITZ around steel reinforcing bars on the durability and bond in reinforced concrete. The papers presented are a mix of theoretical and experimental reports, providing basic concepts as well as identifying conditions and systems where ITZ characteristics affect the overall properties and instance where its influence is small and need not be taken into account in the design of the material and system.

I wish to take this opportunity to thank Dr. A. Katz, the secretary of this conference for his dedication an efforts to bring this volume of papers together, and Professor G. Arliguie, the Chair of committee TC-TPZ and Professor M. Alexander the Co-Chair of TC-ETC for their support and cooperation.

Professor Arnon Bentur
Haifa, March 1998

PART ONE
KEYNOTE LECTURE

1   THE INTERFACIAL TRANSITION ZONE: REALITY OR MYTH?

S. DIAMOND and J. HUANG
Purdue University, West Lafayette, IN, USA

ABSTRACT

The existence of an interfacial transition zone (ITZ) in concrete, and a belief that its existence significantly influences concrete properties have been fundamental tenets in concrete technology for many years. The results of our microstructural investigations carried out over several years on laboratory mixed concretes confirm the physical existence of the ITZ in these concretes, but suggest that major modification in the details of the accepted picture are required.

The presumed basis for the development of the ITZ is a local deficiency in content of unhydrated cement grains near the aggregate interface, attributed to the wall effect. This aspect of the ITZ picture is evident and strongly displayed in all of the young concretes examined. However, the deficiency is statistical and many cement grains close to or touching the aggregate are shown to be present.

The supposed consequence of the statistical exclusion of cement grains from the vicinity of the aggregate is the development of uniform 'shells' of progressively higher porosity and calcium hydroxide contents as the aggregate grains are approached from some particular distance at which the ITZ merges into the bulk cement paste. These consequences are not confirmed. The individual shells are not uniform. Rather, in traverses around the perimeter of a given grain, the actual contents of the components within a shell is highly variable; for example, sampling units of high porosity are found adjacent to sampling units of low porosity. This local variability also holds for the deficiency in unhydrated cement, as well as for the excess in calcium hydroxide. Amalogous local variability is found within the bulk paste as well. Within the ITZ, the variability within a given shell may far exceed the secular change in property with distance from the aggregate. Nor is the 'width' of the altered ITZ the same for its constituent components; we find a 'wide' ITZ with respect to unhydrated cement, a narrower ITZ with respect to pore content, and the calcium hydroxide excess is confined to 10 or 20 μm from the surface.

The magnitude of the excess in mean porosity found in these investigations is much less than that previously reported by others, especially in the ITZ shell closest to the aggregate. In some areas we find some local excesses of porosity at the interface itself, which are attributable to failure of the paste to conform fully to the local aggregate surface, i.e. local failure to fully establish bond.

The Interfacial Transition Zone in Cementitious Composites, edited by A. Katz, A. Bentur,
M. Alexander and G. Arliguie. Published in 1998 by E & FN Spon, 11 New Fetter Lane,
London EC4P 4EE, UK, ISBN: 0 419 24310 0

In consequence of the patchy character of the ITZ and the modest excesses in mean values of the properties measured, it appears that models that predict significant changes in concrete properties with the degree of overlap of adjacent ITZs may not be well-based. Furthermore, the microstructural similarities between paste in the ITZ and bulk paste also suggest that defining the ITZ as a separate 'phase' in concrete is something of an exaggeration.

INTRODUCTION

The idea of an interfacial transition zone or "aureole de transition" in concrete derives originally from optical microscope observations made many years ago by Farran (1). The concept has subsequently been elaborated by a host of authors, using various techniques. Over the last twenty-five years the idea that the zone of cement paste "near" an aggregate interface is somehow microstructurally different from the cement paste more distant from that interface has become the accepted paradigm in concrete technology. An excellent review of this conventional view, and of the evidence on which it is based has been provided by Ollivier, et. al (2).

This difference in structure is assumed by many investigators to have significant consequences on the mechanical behavior of the concrete. A number of so-called 3-phase models have been produced, for example that of Nilsen and Monteiro (3), in which the cement paste in the ITZ is accorded the status of a separate and distinct phase from the rest of the cement paste, and assigned different properties.

It is generally considered that the ITZ cement paste has a significantly higher porosity and is weaker than paste in the same concrete further away from aggregates. An exception to this statement is deemed to apply for high performance concretes produced with the aid of silica fume and superplasticizer. Such concretes are considered to owe their superior properties in major part to the lack of an ITZ.

It is commonly considered that "aggregates" in the ITZ context include sand grains as well as larger coarse aggregate pieces. This is an important factor, since paste contact with sand grains is so much more frequent in concrete than paste contact with coarse aggregate particles.

In considering that cement paste within the ITZ is significantly more porous than bulk cement paste, it is usual to postulate that its local permeability is also greater than that of bulk paste. It follows that overlap in three-dimensional space of permeable ITZ shells of porous paste might provide an easy continuous path for either mass flow of water or diffusion of ions through concrete. This idea of "percolation" through overlapping ITZs has received some support both in terms of model development and in interpretations based on various indirect experimental measurements.

It is evident from many experimental observations that the ITZ concept does in fact have a basis in reality. However, it appears to us that many of the assumptions of the nature of ITZ in real concrete are oversimplified, overestimate the magnitude of differences that may exist between ITZ paste and bulk paste, and fail to appreciate the local variations in microstructure within and without the ITZ. The aim of this paper is to attempt to provide a more realistic working picture of the microstructure of cement paste in concrete, both near and well away from aggregate interfaces.

BRIEF SUMMARY OF THE ACCEPTED FEATURES OF THE ITZ

The generally accepted features of cement paste in the ITZ that are presumed to differentiate it from cement paste in the rest of the concrete may be summarized as follows. The ITZ paste is considered to be different from bulk paste in the same concrete by having (a) a significantly greater content of capillary pore space, much of it interconnected, and thus a higher permeability, (b) a reduced content of unhydrated cement grains, (c) a greater content of calcium hydroxide with a substantial orientation of the calcium hydroxide parallel to the local surface of the aggregate, (d) a greater content of ettringite, and (e) significantly lower strength and stiffness. The ITZ zone is usually indicated as aureole extending $30 - 50$ μm away from the aggregate, the estimate usually being based on microscopic observations. However, as pointed out by Scrivener and Nemati (4), an arbitrary plane sliced through an aggregate is ordinarily inclined to the corresponding plane that is normal to the aggregate surface. Thus distances to some feature measured microscopically on such an arbitrary inclined plane may overestimate the true distance somewhat. In this work all distances will be recorded as they are measured with no attempt at correction for this effect.

In most detailed studies of the ITZ, the properties measured or modeled are scaled along a gradient extending from the actual interface to some distance at which the property ceases to be different from that of the remainder of the cement paste. It is ordinarily assumed that within this zone the gradient of properties is uniformly exhibited around the perimeter of the aggregate. In other words, within the aureole all of the paste at any given distance from the aggregate is considered to display the ITZ parameters characteristics of that distance, to the same extent. In some treatments, the gradient idea is abandoned as an unnecessary complication, and all of the cement paste lying sufficiently close to the aggregate grain to be in the ITZ is accorded or assigned the same property level.

In the following sections we provide the results of ongoing backscatter SEM studies of concrete aimed at elucidating the reasonableness or lack thereof of these working concepts or models.

FEATURES OF CONCRETE AS IMAGED IN BACKSCATTER SEM

Figure 1 and the immediately succeeding figures represent an attempt to convey an appreciation of the nature of hardened cement paste in concrete within and without the conventional ITZ. Figure 1 shows an SEM micrograph of a representative area of a laboratory mixed concrete, taken in the usual backscatter mode at 100x. The specific concrete involved was 3 days old, and was prepared with a normal ASTM Type I portland cement at a water:cement ratio of 0.50. The coarse aggregate and sand were both crushed limestone. The maximum aggregate size was ½ in., and both coarse and fine aggregate gradations were near the middle of the size distribution envelopes recommended for concrete in ASTM C-33. No air entraining agent was used. This concrete was part of a study on effects of mixing, and represented the results of very brief mixing. It was stored in a sealed container for 3 days at room temperature (23° C). These details are provided for the record; the features that will be pointed out are found to a considerable extent in any plain portland cement concrete.

All of the large dark grains that cover most of the field are sand grains. At this low magnification a single coarse aggregate piece would far exceed the dimensions of the entire field. Most of the space between the sand grains is occupied by the obviously composite material we conventionally described as 'hardened cement paste'.

It is of interest at this point to call attention to the range of distances of separation between individual sand grains, i.e. the range of widths of the cement paste layers between adjacent aggregates, as seen in two dimensions on the arbitrary plane depicted. At its narrowest point, the distance between the long sand grain on the left and the sand grain almost spanning the top of the figure is about 25 µm. Corresponding distances between many of the other sand grains are in the 30 to 50 µm range. At the other extreme, the distance between the sand grain spanning most of the upper portion of the field and its neighbor in the lower right corner is nearly 400 µm.

If the zones of hardened cement paste accorded ITZ status were taken as encompassing 50 µm-thick rings around each sand grain, it is evident that many of the ITZ rings would not develop fully because of the closeness of adjacent sand grains. Between other grains the ITZ, as so defined, would encompass nearly all of the paste. In this very ordinary concrete, the non-ITZ or 'bulk' hardened cement paste would actually be present only in more or less isolated patches between particularly widely-spaced aggregates. This element of the microstructure of concrete were pointed out many years ago by Diamond, Mindess and Lovell (5), but seems not be widely appreciated.

The mathematics governing the proportion of ITZ vs. non-ITZ paste for various postulated ITZ thicknesses and aggregate grain size distributions have been elegantly developed in a recent paper by Garboczi and Bentz (6).

Sand grains are obviously the largest features displayed in Figure 1. The next largest features are air voids, few in number in this non air-entrained concrete. Air voids exist within the inter-aggregate space primarily occupied by the hardened cement paste, but obviously do not share the characteristics of the paste. Some of air voids seen in Figure 1 are very close to sand grains, and thus fall partly or completely within the ITZ as geometrically defined. Nevertheless, they are generally ignored in considering or modeling ITZ phenomena. While the air voids in this concrete are sparse, in ordinary air-entrained concrete the air void volume often constitutes 12 to 15% of the total paste + air void volume, a feature that should not be ignored.

The next smallest but still obvious feature of Figure 1 is the bright/dark texture of the hardened cement paste composite. The bright grains obviously represent individual residual unhydrated portland cement grains, and range in size up to about 80 µm. The relatively large proportion of unhydrated cement grains overall is a consequence of the limited hydration in this 3-day old concrete. It should be pointed out that these unhydrated cement grains are not as homogeneous as they appear in the figure. The gray scale distribution used for Figure 1, smears out the significant difference in brightness between ferrite and C_3S, and the lesser differences in brightness between the C_3S, C_2S, and C_3A components of the grains.

The remainder of the hardened cement paste composite appears in shades of gray or black. The brighter gray areas represent calcium hydroxide; the darker gray areas are primarily C-S-H, but include any ettringite or monosulfate present. The capillary pores, not generally visible at the magnification of Figure 1, are effectively black.

A careful attempt at a visual assessment of possible difference in microstructure between the conventionally-defined ITZ zone and bulk paste further away from the aggregate in Figure 1 might show some differences. However, at this scale the differences are not obvious, and clearly there are no sharp boundaries marking off 'ITZ paste' from 'bulk paste'.

One additional point in Figure 1 is worth noting. It can be in Figure 1 that a number of unhydrated cement grains touch or almost touch the aggregate particles, i.e. they occur well within the zone normally considered to constitute the ITZ. It is evident that cement particles are not excluded from the vicinity of aggregate surfaces on any absolute basis.

Many features of the concrete microstructure require higher magnification to be appropriately visible. Accordingly, Figures 2, 3, and 4 are higher magnification micrographs taken of specific areas within Figure 1 so as to illustrate some of these features.

Figure 2 is taken from the lower left hand portion of Figure 1 and shows a part of the long sand grain visible there. It is seen that a section of the perimeter of the sand grain is covered with an irregular bright gray deposit, which is calcium hydroxide. Similar coverage by calcium hydroxide of portions of the perimeter of various sand grains can be seen by careful examination in Figure 1. Occasional deposits of calcium hydroxide on aggregate surfaces are characteristic features of most concretes. The degree to which such deposits should be considered as falling within the ITZ is not necessarily clear. They are in some sense not deposited within the paste, but rather deposited on the aggregate.

Above and below the calcium hydroxide deposit on the sand grain in Figure 2 are local regions or patches of paste of higher than usual porosity, as indicated by the arrows. Similar local patches of high porosity occur irregularly along aggregate surfaces in most concretes. Such patches of high porosity are also found in bulk paste far removed from aggregate surfaces.

Further details are visible in Figure 3, taken of the same area at a still higher magnification (1000x). The irregular nature of the calcium hydroxide deposit is clear, as is the fact that there is a partial disconnection of the deposit to the underlying sand grain. Figure 3 also shows a number of additional features.   The coarse and presumably interconnected character of the pores within the ITZ region, and also in the bulk paste far removed from the aggregate is evident. Several small hollow-shell hydration grains can be seen near the right hand margin of the figure. A number of "partly hydrated phenograins" consisting of dense gray 'inner product' hydration shells surrounding bright residual unhydrated cement grains are also clearly visible.

Figure 4 represents the area around the small triangular sand grain to the left of the center of Figure 1. A narrow deposit of calcium hydroxide is observed near the rightmost apex of the triangular sand grain, but most of he rest of the perimeter of the sand grain is free of such deposits. A patch of high local porosity occurs adjacent to the middle portion of the lower surface of the grain. Other local patches of high porosity can be seen in the bulk paste below and distant from the grain.

Figure 4 contains portions of two other sand grains on either side of the triangular grain, with narrow channels of paste separating them. The channels are each about 50 – 60 μm wide. Attention is called to the fact that a significant content of unhydrated cement grains occurs within these narrow channels, despite the fact that all of the paste in the channels would be entirely within the ITZ as ordinarily defined.

Figure 5, the last of this set, shows the region around the sand grain in the lower right corner of Figure 1 at higher magnification. It is seen that the entire left-hand perimeter of the sand grain is separated from the adjacent hardened cement paste by a straight-sided crack about 3 μm wide. The crack exactly follows the perimeter of the sand grain, then projects irregularly into the paste above it and tapers to a vanishing point just at the top of the figure. This is obviously a bond crack, a very different feature from a porous zone within the ITZ paste. Bond cracks were found rarely in well-mixed concretes, but slightly more commonly in brief-mixed concretes such as is pictured here. Some bond cracks may be artifacts of drying and exposure of the sample to high vacuum, but similar cracks are sometimes observed in optical microscopy where no vacuum has been applied. They are interface features, not interfacial zone paste features; and clearly should not be included in tallies of interfacial zone pore space.

The general picture of the microstructure of the concrete presented in Figures 1-5 are reasonably representative for portland cement concretes in general. Of course, specific details may vary. Concretes batched at lower water:cement ratios tend to have fewer local patches of high porosity and the paste is generally less porous. Older concretes have fewer and smaller residual unhydrated cement grains. Air-entrained concretes have many more air voids. Richer concretes and those with coarser sand particles show greater distances of separation between adjacent aggregate particles. As will be seen later, even the degree of mixing affects the microstructure; prolonged mixing produces a less coarsely porous paste than is seen here. Superplasticized, silica fume bearing, and fly ash bearing concretes each show characteristic microstructural modifications.

QUANTITATIVE ASSESSMENT OF THE ITZ BY IMAGE ANALYSIS

Image analysis of backscatter SEM micrographs is an appropriate procedure (within limits) for quantitatively assessing the spatial characteristics of the hardened cement paste component of concrete. In such measurements the usual procedure is to quantify variations of specific features in successive narrow bands starting from the actual interface and extending outward to bulk paste. This type of analysis was pioneered by Scrivener and Pratt (7). These authors analyzed successive 5 μm wide bands, the object being to provide an overall statistical evaluation of the variation in area% of pores, calcium hydroxide, and other features, with distance from the surface of the aggregate.

Our image analysis procedure is somewhat different from that of Scrivener and Pratt, partly because of differences in image analysis program capability. An explanation is in order.

Each SEM micrograph captured for image analysis is actually a computer file representing the array of brightness levels at each of the pixel points constituting the image. Subdividing the image into successive strips from the aggregate surface for analysis is not possible with our image analysis program unless each strip is individually separated from the remainder of the file. This is done using the "Cut" command in Photoshop™. In our work we have standardized on a 10 μm strip width; narrower strips are impractical.

However, examination of, for example, Figure 3 suggests that the inherent scale of the individual units (hydrated particles, pores, residual unhydrated grains) are quite commensurate with a 10 μm strip width. This is in accord with the results of earlier work on paste by Wang and Diamond (8) who made measurements the average sizes of the microstructural features in hardened cement paste. For w:c 0.45 pastes, the mean sizes found in that work were slightly more than 2 μm for pores, approximately 7 μm for calcium hydroxide, and about 11 μm for unhydrated cement grains.

For technical and practical reasons, most backscatter SEM image analyses are done at a standard magnification of 500x. At this magnification the area imaged is about 250 μm across. Only a part of the perimeter of an individual sand grain and its surrounding paste can be included at this magnification. To get a complete picture of the paste surrounding any one sand grain of appreciable size requires a traverse around the grain consisting of a number of adjacent 500x micrographs.

We have noticed wide variations in features around any given sand grain in conducting such traverses. We consider that this lateral variation constitutes important microstructural information that should not be lost in sampling and averaging. Accordingly, in all of our analyses a succession of adjacent images are taken all of the way around each sand grain chosen for evaluation. To avoid overlap a slight gap is left between adjacent images.

As indicated previousy, each image is approximately 250 μm long. In carrying out the analyses for variation with distance from the sand grain interface successive 10 μm-wide strips are cut from the image and analyzed separately. Thus the primary sampling unit in our analysis is a narrow strip of hardened cement paste approximately 250 μm long and 10 μm wide, following the contour of the aggregate, and positioned at a definite location around the traverse and a specified distance from sand grain surface.

Our analysis, like that of most others, is carried out at an image resolution of 512 pixels, and as stated previously, a magnification of 500x . Under these parameters the size of each individual pixel is approximately 0.5 μm x 0.5 μm. Thus the primary sampling unit contains approximately 10,000 individual pixels. A full traverse around a large sand grain might consist of 70 or 80 primary sampling units.

Each pixel of each sampling unit undergoes a binary segmentation on gray level and is assigned as follows: the brightest pixels are allocated to unhydrated cement; the next brightest to calcium hydroxide, and the nearly black pixels are assigned to pore space. The remaining pixels of intermediate gray level are assigned to a leftover "hydrated cement" category. Such pixels include areas of C-S-H (both 'inner product' or 'hydrated phenograin' C-S-H and the 'groundmass' C-S-H), and also included ettringite, monosulfate, and probably much pore space finer than the size of an individual pixel.

The segmentation is not perfect. Unhydrated cement pixels are very easily and reliably distinguished from other phases. Calcium hydroxide is easily separated from unhydrated cement, but its distinction from 'hydrated cement', i.e. C-S-H, has some element of arbitrariness. "Pore" pixels are conveniently distinguished from pixels of any solid component with the help of a paint program. Such a program permits easily visualization of the effects of adjustment of the boundary level so that one can accurately assess the gray level boundary that best delimits the pore spaces.

An idea of how the whole procedure works in practice can be obtained from Figure 6, which illustrates the results of processing two adjacent primary sampling strips around a portion of a sand grain. The specific concrete was a 3-day old, brief-mixed quartzite aggregate concrete of 0.5 w:c ratio. Each sampling units is 10 μm wide and about 250 μm long, and is cut parallel to the perimeter of a sand grain. The particular units depicted are respectively 40-50μm and 50-60 μm away from the perimeter of the sand grain.

In Figure 6 the 'A' pair of images display the original unsegmented backscattered image containing all of the pixels at their original brightness. Each of the 'B' pair represent the contents of a modified file derived from the corresponding 'A' file in which all pixels except those representing pore space have been removed, and the pore pixels have been painted an arbitrary color. The 'C' pair represent calcium hydroxide locations, and the 'D' pair show the pixels assigned to unhydrated cement. All pixels not assigned to the 'B', 'C', or 'D' categories are assigned to the 'hydrated cement' category, and were not separately plotted.

In further processing, the area% of pixels assigned to each category are calculated for each primary sampling unit.

As will be shown later, large variations occur in the area% values found for a given category within a single strip as we process successive images around a sand grain at a given distance from it. This appears to be a consequence of real fluctuations of paste structure, a feature that has not been commented on to any great extent in previous work.

In compiling our overall ITZ assessment for a given concrete, we combined average results, strip by strip, around several randomly selected sand grains. Because of the size of the sand grain relative to the paste structural units and the variations around the grains, we find that averaging results around four randomly chosen sand grains provides effective and repeatable characterization.

Three complications should be discussed prior to presenting these results. First, it is not possible to maintain the same number of successive sampling units as we move around the perimeter of the aggregate. In some directions an adjacent sand grain interposes after only a short distance; in other directions, the next sand grain may be comparatively far away. Successive sampling units that represent an attempt at measuring the gradient within the ITZ can obviously only be taken only half way to the next grain. Thus, as we compile tallies of third, fourth, fifth, sixth, and subsequent successive strips, sampling units are necessarily missing in some directions, and the number of individual units available to be averaged for the more distant strips progressively decreases.

The second feature involves the complication induced by the occasional bond crack, such as that seen in Figure 5. Where such a crack exists, the origin of the first paste strip is taken at the outer surface of the hardened cement paste, not the surface of the aggregate. This eliminates the crack pixels from the tally.

The third complication has to do with air voids. As previously mentioned, in our view air voids do not constitute a portion of the hardened cement paste. To avoid any biasing of pore space measurements, all sampling units that contain air void pixels were simply eliminated from the averaging process.

RESULTS

We have examined a variety of concrete specimens using the technique described above. Major variations included the chemical character of the aggregate (limestone vs. quartzite), the degree of mixing, and the age. The end members with respect to degree of mixing were brief mixing for 30 seconds (with an efficient mixer) on one hand, and a full 10 minutes of mixing followed by modest vibratory consolidation on the other. Specimens were taken at 3 days to represent the early-age microstructure, and at 100 days to assess the effects of the additional hydration. We have examined paste areas around sand grains and around coarse aggregate particles, and have cut sections horizontally and vertically to assess gravitational field effects. This work is still in progress, and we report here only a portion of the results

The results of each analysis for mean value of the parameter concerned vs. distance from the aggregate are presented in a series of bar plots below.

Analyses for Well- Mixed, 3-day Old Concrete Containing Crushed QuartziteAgregate

The set of analyses obtained for well-mixed 3-day old concrete made with quartzite sand and quartzite coarse aggregate will be used as a pattern for subsequent comparisons.

Figure 7 provides a high magnification illustration of the paste microstructure in this particular concrete, for comparison with Figure 3. The difference in texture of the groundmass of the hardened cement paste composite is evident, and will be apparent in a quantitative comparison of pore area% for the two concretes to be provided later. Figure 7 also shows a bond crack, and several cracks appear in the sand grain itself.

Figure 8 is a simple bar plot depicting the average area% of unhydrated cement grains found in successive strips starting from the aggregate surface. It provides striking confirmation of the basic assumption underlying the ITZ paradigm; that the wall effect limits the content of cement grains near the aggregate surface. We have found patterns similar to that of Figure 8 in all of the concretes we have examined.

We have previously pointed out instances of the occurrence of unhydrated cement grains close to aggregate surfaces. The data of Figure 8 do not conflict with this. Unhydrated cement grains do indeed occur near aggregates; the data of Figure 8 indicate that they occur near aggregates at a frequency significantly less than they do elsewhere in the paste.

The gradient depicted in Figure 8 is steep. A general level of about 16 area% of unhydrated cement grain pixels is found in the strips beyond 50 μm. Only about 4 area% of unhydrated cement pixels is found on average in the innermost 10 μm strip. There is a clear gradient in between. This very pronounced gradient effect serves as a very clear marker for the existence of an ITZ.

However, it is not the shortage of residual unhydrated cement grains, but rather the increased content of pores that is commonly taken to be its most important characteristic of ITZs. Figure 9 displays the averaged results for area% of pore pixels in the same format as Figure 8. It is seen that in this concrete the "bulk" paste averages about 7 area% of detectable pores. This value rises to 10 or 11area% within the major part of the ITZ, and to 13 area% in the innermost 10μm strip. In considering these values, it should be recalled that pores much smaller than the size of a pixel will not be tallied.

Figure 10 shows the corresponding results for calcium hydroxide. Here only the innermost 10 μm-wide strip shows an indication of excess calcium hydroxide, showing about 17 area% as compared to 12 to 15area% for the rest of the paste - both within the conventional ITZ region and in the bulk paste. The occurrence of calcium hydroxide deposited directly on aggregate surfaces has been illustrated and commented on earlier.

Finally, Figure 11 shows that on average, approximately 2/3 of the area of all of paste in this concrete is allocated to our generalized 'hydrated cement' category, i.e. is not distinguished as either unhydrated cement, pore, or calcium hydroxide. There is clearly no significant change in this parameter with distance from the aggregate; i.e. 'ITZ' paste and 'bulk' paste are identical in this regard.

Effects of Degree of Mixing on Interfacial Zone Parameters

In the next group of figures we explore the effect of extremes in the degree of mixing of the concrete on the numerical parameters exhibited by the various ITZ features. In them we compare the results obtained above with those for otherwise identical concretes mixed very briefly, but sufficiently to produce grossly homogeneous concrete.

Figure 12 shows a split bar chart with the data for unhydrated cement grain area% for brief-mixed (30 seconds) quartzite aggregate concrete compared with the results for corresponding well-mixed concrete previously presented. It is seen that the gradient in this parameter, attributable to the wall effect, is essentially the same for brief mixing as it is for thorough mixing. It appears that the wall effect not a feature that can be overcome by prolonged mixing.

In Figure 13 an unexpected effect emerges. Figure 13 shows a split bar comparison chart for pore area%. It is seen that the area% of detectable pores, unlike the area% of unhydrated cement grains, is influenced by the extent of mixing. Indeed the effect is profound, in both the 'ITZ' paste and the 'bulk' paste.

Specifically, the concrete produced by brief mixing has, generally speaking, about twice the content of detectable pores as the well-mixed concrete. Figure 13 indicates that for the brief-mixed concrete the average area% of pore pixels in the bulk paste is around 17%: it rises only to about 21% in the innermost ITZ strip. We have found similar results for the analogous brief-mixed limestone aggregate concrete. An indication of the porous character of all of the paste, ITZ and bulk, in such concrete is visible in Figure 3. It appears that concretes mixed only briefly develop coarsely-porous paste; more extensive mixing significantly reduces the content of coarse pores.

The proportional excess of pore space in the ITZ paste over that of the bulk paste in such brief-mixed concretes appears to be of little consequence, since all of the paste is so highly porous and the pores are obviously interconnected in the bulk region as well as in the ITZ. It is notable from the data of Figure 13 that the area% of pores in the bulk paste of the brief-mixed concrete is much higher than the paste in even the innermost strip of the ITZ in the corresponding well-mixed concrete.

Figure 14 provides the analogous comparison for calcium hydroxide. In the brief-mixed concrete calcium hydroxide contents above the level present in the bulk paste was found in both of first two 10-μm wide strips. In the well-mixed concrete the excess was confined to the innermost strip. The wider spread within the ITZ of the brief-mixed concrete may represents extra local deposition facilitated by the larger open spaces in the paste.

Finally, Figure 15 provides the analogous comparison for effect of mixing on the residual 'hydrated cement' component. The general level found for the brief-mixed concrete is only about 51area%; this is somewhat less than more than 60 area% in the corresponding well-mixed concrete. Again the ITZ paste and the bulk paste show identical amounts, i.e. no variation is found with distance from the aggregate. This is a quite remarkable result.

Effect of Aging on Measured Interfacial Zone Parameters

In Figures 16-19 we present split-bar graphs illustrating the observed effects of aging on the measured parameters of the well-mixed quartzite-bearing concrete previously examined. The comparison is between 3-day old concrete, previously discussed and 100-day old concrete. The latter age is taken to represent effectively mature condition for the concrete.

Figure 16 shows the effect of aging on the content of unhydrated cement. As expected, the additional hydration experienced in the 97 days of extra aging reduces the content of unhydrated cement very substantially. However the plot shows that some evidence for the characteristic gradient is still discernable even at 100 days.

The effect of aging on pore area% is shown in Figure 17. The additional hydration has resulted in some reduction in porosity, as expected, but the decrease in porosity is modest. The change is more or less uniform with distance from the aggregate, and some modest extra pore space continues to be detected in the ITZ.

Figure 18 indicates that despite the decrease in the content of unhydrated cement, little if any detectable increase in the area% recorded for calcium hydroxide results from the additional aging. The apparent *decrease* with time of calcium hydroxide in the 0 − 10 μm strip, from 17 area% to 15 area%, is presumably a statistical anomaly.

Finally, Figure 19 shows an expected increase in general area% of 'hydrated cement' with aging, from about 62area% to about 75area%. Again the hydrated cement component appears to be uniformly distributed across the field and does not vary with distance from the aggregate.

Comparison of Sand Grain and Coarse Aggregate Grain ITZ Parameters

All of the results so far displayed were compiled for the hydrated cement paste around and near sand grains. The question naturally arises as to whether similar results would be obtained for corresponding compilations around coarse aggregate pieces. We have examined this question in both brief-mixed and well-mixed limestone-bearing concretes.

Securing adjacent sampling units at the requisite 500x magnification entirely around a coarse aggregate grain is a daunting task. Almost 100 adjacent micrographs are required to secure a complete traverse around a single one-half inch coarse aggregate piece. Nearly 500 individual sampling units need to be individually evaluated for each coarse aggregate grain.

This work is still underway. Visual observations and results to date suggest that there are no real differences between the distributions of the different cement paste components around coarse aggregate pieces and around sand grains. In particular, we have established that, for well-mixed limestone bearing concrete, the area% of unhydrated cement, of pores, and of 'hydrated cement' around coarse aggregate particles and around sand grains have the same general levels and similar trends with distance from the surface of the aggregate. Both general levels and trends with distance are similar to those shown in Figures 7, 8, and 10, for investigations around sand grains in quartzite-bearing concrete. However, it appears tentatively that the area% of calcium hydroxide surrounding coarse aggregate particles may be slightly smaller than for sand grains, both within the ITZ and further away.

Local Variations Around Sand Grains and Their Significance

The analyses so far described were formulated in terms of overall statistical average of area% for each characteristic (e.g. pores, calcium hydroxide, etc.) with distance from the aggregate surface. This is the classic approach to microstructural investigations of ITZ characteristics. Unfortunately, the averaging process hides a most important feature, the neglect of which we believe to have resulted in an incorrect picture of the ITZ, and which may invalidate some of the current ITZ models and applications.

The concept involved is perfectly straightforward. Cement paste is an inherently patchy material, and as is evident in the micrographs presented earlier, this patchy character is as characteristic of the region of paste included in the ITZ as it is of the bulk paste.

In the usual concern with the ITZ attention is centered on the changes in *average* levels of parameters with distance from the interface. There is a stated (or unstated) assumption that all of the paste in a particular "shell" at a particular distance from the interface has the same value of the parameter. This is an appropriate assumption if departures from the average values are small, but unrealistic if departures are large. In this section we explore the degree to which this assumption is realistic.

To illustrate the results of our explorations of this point we have chosen data secured around a single sand grain taken from the data set for the well-mixed quartzite aggregate concrete hydrated for three days, the concrete of Figures 1-5. We have examined similar data sets for the other concretes and find that the degree of variability illustrated here is more or less characteristic of the entire spectrum of concretes examined.

Our procedure of collecting data from laterally adjacent sampling units so as to complete a circuit entirely around a sand grain has been described previously. For the particular small sand grain selected, eight adjacent sampling units were required to complete the circuit. We have used a different bar chart format in the following figures to emphasize that what is being compared here is not overall average values vs. distance from the aggregate, but rather individual variations found in adjacent sampling units all at a constant distance from the aggregate.   The particular distance is specified in each example.

The ITZ parameter most often considered is porosity; accordingly, our illustration will concentrate on this parameter.

For the sand grain in question, the mean value of area% pore space for the innermost 10 μm strip is 8.13%. Figure 20 provides a bar chart to illustrate graphically the area% pore space to be expected under the hypothetical assumption that each of the eight adjacent sampling units making up this strip had the same content of pores.

The actual data for the eight individual sampling units making up the strip is presented in bar chart form in Figure 21. The pore contents are far from uniform from sampling unit to sampling unit. In fact the individual sampling units range in pore content detected from a low of 4 area% to a high of 13 area%, in more or less random fashion.

The obvious question is to what extent the variations in local area% of pores depicted in Figure 21 is real variation, as contrasted with variation associated with the statistics of assessing the value.

As described earlier, each sampling unit encompasses approximately 10,000 individual pixels, each individually assigned to one category or another by the binary segmentation process.   In a primary sampling unit having 8.13 area% pores, approximately 800 of the 10,000 pixels have been assigned to the pore category. The statistical variation of this process is analogous to that of counting statistics for x-ray peaks, treated many years ago by Klug and Alexander (9) in their standard reference.

According to the Klug and Alexander treatment, the relative standard deviation (i.e. the coefficient of variation) for a given counting process is inversely proportional to the square root of the number of counts tallied, i.e.

$$U_s = 1/\sqrt{N}$$ (Eqn. 1)

Where: U_s is the relative standard deviation of the measurement and
N is the number of counts secured.

It is understood that this statistic reflects only the component of variation stemming from the counting process itself, with a homogeneous population assumed to be present. Any actual variations within the population naturally increase the fluctuations due to the counting process.

For a measurement consisting of 800 counts, the relative standard deviation is 2.7%.

Variations from sampling unit to sampling unit on the scale illustrated in Figure 21 are clearly far beyond the modest statistical variation expected. Thus significant real local variation in area% pore space exists in adjacent sampling units. This data reflects the presence in adjacent sampling units of patches of relatively high porosity and patches of relatively low porosity, in the strip closest to the aggregate.

Similar variations are also found within the 10 μm-wide strips further away from the interface. The lateral variations within a strip far exceed the 'secular' reduction in area% pores with distance from the aggregate surface, depicted for example in Figure 17.

In examining variations on a strip-by-strip basis, the further outward we go, the more incomplete the tally becomes. In some directions we reach the half-way point to the next aggregate, and necessarily stop collecting primary sampling units. By the time we reach the 6th through 9th strips that represent paste beyond the range of the conventional ITZ, sampling units can be taken only in a few directions. In this data set only three sampling units each are available in each of the 6th and 7th strips, and only two each in the 8th and 9th strips. Thus in the bulk paste our ability to assess the pattern of variation of *adjacent* sampling units is necessarily lost. However, as illustrated in Figure 22, it is possible to assess the general extent of variation of these bulk paste units. In Figure 22 the first 3 bars represent non-adjacent units in the 6th strip; the next 3 bars represent non-adjacent units in the 7th strip; and the last 4 bars represent non-adjacent units in the 8th and 9th strips, respectively.

The overall variation in the "bulk" paste is clearly of the same order as that of the paste in the ITZ, and in our opinion both represent an inherent characteristic of hardened cement paste.

The variation among adjacent sampling units within the conventional ITZ for the other parameters measured are also of interest. It appears that variation of calcium hydroxide contents within a particular strip in the ITZ even exceeds the variation of pore contents. Figure 23 shows the area% calcium hydroxide in adjacent sampling units within the 0 – 10 μm strip whose pore area% variation was plotted in Figure 21. Calcium hydroxide area% values in the individual sampling units range from 5% to over 40% *in the same strip*! Examination of the SEM micrographs of Figures 3 – 5 clearly show the basis for this effect. Many areas adjacent to the sand grain are virtually free of visible calcium hydroxide, but in some areas obvious deposits are seen attached to the aggregate surface. This localized visible deposition of calcium hydroxide on patches of aggregate surface is a common feature to all of the concretes we have examined.

It appears that the bulk paste has much less variation in calcium hydroxide content than the near-surface strips. Figure 24 shows the area% calcium hydroxide for the same 10 non-adjacent sampling units whose pore contents were plotted in Figure 22. It seem that in the bulk paste the local calcium hydroxide contents within sampling unite vary only between about 10 area% and 15 area%.

Earlier, we indicated that the ITZ is marked by a steep and characteristic gradient in the average area% of unhydrated cement grains with distance away from the interface, as shown for example in Figures 8 and 12. It is of interest to examine whether this feature, with such a marked trend in average value with distance from the interface, is also subject to significant lateral variation within any given strip. Figure 25 suggest that it is. The average area% for the innermost strip around this sand grain is only about 5 %. However, values for individual sampling units within the strip range from as little as 2% to as much as 11%.

The bulk cement paste has a much higher average content of unhydrated grains, than the first ITZ strip; it appear also that bulk paste shows a much greater range of variation in this property. In Figure 26 two of the 10 bulk sampling units that constitutes the set have only about 5 area% of unhydrated cement; two others have 30 area% or more.

DISCUSSION

The answer to the question posed by the somewhat provocative title assigned to this paper by the Conference Organizing Committee is straightforward: the ITZ exists, and it is *not* mythical. Our investigations nevertheless suggest a considerably different picture of the ITZ than is usually postulated.

Before proceeding to these concerns, it seems appropriate to compare the general trends of the overall data obtained here to those previously reported by other investigators.

The general trends of the variation of the mean area percentages of unhydrated cement grains and calcium hydroxide with distance from the aggregate reported here fairly closely follow those previously reported by Scrivener and various co-workers. For example, Scrivener, Bentur, and Pratt (10) found the mean unhydrated cement content to fall from approximately 25% in a strip 30 − 40μm from the aggregate surface to about 4% in the 0-10 μm strip, in a 1-day old plain concrete. Scrivener, Crumbie, and Pratt (11) reported mean unhydrated cement contents to fall from approximately 15% at 50 μm to approximately 4% at 10 μm (and to even smaller values closer to the interface). This strong trend was fully described and its significance explored by these authors.

With respect to calcium hydroxide, Scrivener, Bentur and Pratt (10) reported finding essentially no variation in this parameter with distance from the aggregate, the overall level found being only about 7%. The cement they used contained some slag, which may have limited the general level of calcium hydroxide produced. In analyses of interfacial zones formed around single pieces of aggregate placed in previously-mixed paste (not necessarily identical with concrete) Scrivener and Gartner (12) found mean calcium hydroxide contents mostly within a band ranging from 11% to 16% with the higher values generally within 20 μm of the aggregate. These results are more or less commensurate with the present findings.

This general agreement does not necessarily extend to pore content measurements. Scrivener and co-workers generally reported substantially higher values for this parameter than reported here. For example, in the plain concrete sample reported by Scrivener, Bentur, and Pratt (10) the mean pore content reported was about 20% in both their $20 - 30$ μm and $30 - 40$ μm strips, rising to well over 30% in the $0 - 10$ μm innermost layer. This is far more pore space than found in any of the concretes reported here.

Scrivener, Bentur, and Pratt (10) used 10-μm wide strips in their analyses, providing for easy comparison with the present results. In most other reported work, Scrivener and co-workers provided mean values over 5 μm-widedistances. A noteworthy feature of their results was that the general trend toward increased pore space as the interface was approached accelerated enormously in the last strip, i.e. as the actual interface was approached. For example, Scrivener, Crumbie and Pratt (11) reported a bulk pore space value approximating 10%; the third 5 mm strip had about 13% average pore space, the second about 16% average pore space, and the innermost 0-5 mm strip over 30% average pore space. Similar extremes of elevated pore content within the first 5 μm from the surface were reported by Scrivener and Gartner (11).

These authors plot average pore space vs. distance as smooth curves, or at least as continuous functions. The implication is that the extreme values of pore space found in the innermost 5 mm layer are merely continuations of the lesser increases found within the ITZ layers not in direct contact with the aggregate.

In attempting to assess whether the average pore content might indeed be expected to rise smoothly to such high values in the $0 - 5$ μm zone, the present writers have qualitatively examined a representative set of their own $0 - 10$ μm wide strips, specifically those for well-mixed 3 day-old limestone-bearing concrete. A high magnification SEM of the microstructure of this concrete was provided as Figure 7. The strips specifically examined were those segmented to show only the pore pixels. The idea was to see whether in these 10 μm-wide strips the innermost 5 μm portion had a detectably greater content of pore pixels than the outer 5 μm portion. What we found was that while many of the approximately 250 μm-long strips showed no particular difference in pore content between the inner and outer parts, some did. In the strips that did, the extra pore pixels were present as chains of pixels along and in mostly in contact with the surface of the aggregate. A representative example of each of the two patterns are provided as Figure 27a and Figure 27b.

It appears that these "extra" pore pixels in contact with the aggregate represent local regions of bond failure, i.e. locations at which the cement paste failed to conform closely to the aggregate surface. Such irregular zones are quite different from bond cracks like that shown in Figure 5; in the writers' opinion they are also different from individual pores that constitute characteristic features of <u>within</u> the hardened cement paste. It should be noted that bond cracks *per se* were specifically excluded from measurements reported by Scrivener and Gartner (12) and presumably from the other measurements reported by Scrivener and co-workers as well. Thus the large excesses of pore space found by these authors in the 0 - 5 μm strip do not represent the effects of bond cracks. To what extent they may represent local regions of bond failure as defined here rather than regions of high porosity within the cement paste the present writers can only speculate.

In any event, a rather different picture of the ITZ emerges from our considerations than that commonly accepted. The usual picture is that successive rings or shells of progressively altering properties exists around each aggregate, and that a quite definite gradient in properties exists from the aggregate surface outward to the bulk paste. Each shell is presumed to have uniform properties all around the aggregate. It is also assumed the ITZ has a definite thickness or width, at which all of its properties merge into those of the bulk paste. None of this appears to be true.

It is evident from our results that even where a gradient exists and is strongly marked within the ITZ, as it does for content of unhydrated cement grains, the gradient is statistical in nature. Significant local variation is observed in any particular shell as we go around the aggregate grain. We cannot define a physical gradient along any arbitrary radial line from the surface; an area of high local unhydrated aggregate content can and does exist "inside" of an adjacent area of much lower local content.

This statistical character of the gradient is even more pronounced for pore contents and calcium hydroxide contents, where the statistical mean value gradient is much weaker than it is for unhydrated cement. For pores and calcium hydroxide contents the variation from sampling unit to sampling unit within a particular shell grain seems far more significant than the statistical gradient effect.

One implication from this picture has to do with the concept of percolation as it is applied to the ITZ. It is commonly considered that if the spacing between neighboring aggregate grains is narrow enough that the ITZs of adjacent grains characteristically overlap each other in space, a continuous linked path of high porosity can be produced entirely across the concrete, i.e. the pores "percolate". Our results suggest otherwise. In our measurements the excess content of pores is limited, but more importantly, because the pore space so badly distributed around any particular aggregate. Even local percolation would seem to be restricted to those few instances where a highly porous patch of an ITZ overlaps another highly porous patch of an adjacent ITZ. The probability of such events forming a continuous chain through a concrete member seems very small.

The second characteristic usually assumed for the ITZ is that it has consistent width for all of its altered properties, i.e. there is some distance from the interface at which its properties merge with those of the bulk cement paste. This too appears to be incorrect, even as a statistical effect.

In all of our data, the statistical gradient for mean area% of unhydrated cement vs. distance is clearly defined, and the gradient extends over the first five 10 μm – wide strips, i.e. 50 μm, before this parameter merges with the background level of the bulk paste. The indicated ITZ width for this feature is clearly 50 μm. In contrast, the plots for the average area % of pores typically show little difference beyond the first three 10 μm – wide strips, i.e. the indicated ITZ width seems to be about 30 μm for this feature. It is harder to be definite because the gradient itself is so weak. For calcium hydroxide the excess content found is either confined entirely to the first 0 – 10 μm strip or to the first two strips. Thus there is no single interfacial zone width for the different features.

This being the case, models that define the cement paste in the ITZ as a separate phase in concrete distinct from the bulk cement paste face significant conceptual problems in defining the width of the new phase. Actually, the microstructural similarities of the bulk and ITZ-region pastes as seen in the SEM figures presented earlier, and their common patchy character suggest that such '3-phase' models are not particularly appropriate descriptions of concrete.

Specifically, the patchy character of the distribution of the ITZ parameters laterally at any particular distance from the aggregate appears to be a reflection of the equally patchy character of the distribution of these parameters in the bulk cement paste. Hardened portland cement in concrete is clearly a locally , non-uniform material, with area to area variations accessible to image analysis, and readily visible by inspection of individual SEM micrographs.

An inference from this patchy picture of the ITZ in ordinary concrete is that any mechanical or physical effect associated with different degrees of ITZ overlap may be much weaker than expected. The special picture with respect to percolation has already been mentioned.

That this inference may in fact be correct seems to be supported by experimental results reported in a just completed Ph. D. thesis by Rangaraju (13). Rangaraju prepared a series of concretes in which the aggregate grain size distribution was systematically varied to provide wide range of inter-aggregate spacings. The average inter-aggregate spacings on an arbitrary plane for these concretes were measured by standard linear traverse measurements. The results ranged from 60 μm at one extreme to 160 μm at the other. The concrete with the 60 μm average spacing would have very significant overlap of ITZs as conventionally defined; at the other extreme the 160 μm concrete would be expected to have almost no ITZ overlap. Despite this, rapid chloride permeability test measurements gave almost identical results for all of the specimens of a given w:c ratio, regardless of the mean inter-aggregate spacing. A similar lack of variation with mean inter-aggregate spacing was found for compressive and tensile strength measurements and for measurements of the dynamic modulus of elasticity.

The specific effect that the degree of mixing appears to have on the content of detectable pores deserves comment. We had expected that brief-mixed concrete might have a more porous *interfacial zone*, and that some of the extra pore space would be removed by more extensive mixing. Under this reasoning the marginal effect of extra ITZ porosity would be greater for brief-mixed than for thoroughly mixed concrete. The results indicate that brief-mixed concrete has an extremely high content of detectable pores both in the bulk paste and in the ITZ, and the statistical excess associated with the ITZ is proportionately less significant.

CONCLUSIONS

We draw the following conclusions from the results described here:

1. Hardened cement paste near concrete aggregates has statistically lower contents of unhydrated cement grains and statistically higher contents of pore space and to some extent of calcium hydroxide than hardened cement paste further away from aggregates. To this extent the conventional ITZ concept is valid.

2. The usual ITZ assumptions that (a) each aggregate particle is surrounding by successive shells each homogeneous laterally around the aggregate, and that (b) the successive shells exhibit an actual physical gradient of properties with distance from the aggregate, and that (c) this gradient extends to some particular distance from the aggregate marking the limit of the ITZ, and (d) that at this limit the properties of the ITZ paste merge smoothly with those of the bulk paste are not in fact correct.

3. The paste in a given shell is not homogeneous around the aggregate particle; rather adjacent local areas within a given shell have widely varying contents of unhydrated cement grains, pore space, and calcium hydroxide. The random variation between adjacent areas within a shell is often much greater than the average change in the parameter between successive shells, i.e. with distance from the aggregate.

4. The geographic limit of statistical departure of properties from those of bulk paste, i.e. the 'width' of the interfacial zone, is different for each characteristic measured even within the same concrete. In our concretes ITZ effects extends to 50 μm with respect to deficiency in unhydrated paste content, to ca. 30 μm for excess pore space, and only to 10 or 20 μm for excess calcium hydroxide. Thus no single width applies.

5. The proportion of pixels not specifically allocated to pores, residual unhydrated cement, or calcium hydroxide, here defined as 'hydrated cement', and consisting mostly of C-S-H, is remarkably constant in any concrete. No gradient is exhibited, and the ITZ and bulk paste values are identical.

6. The proportional departure of statistically-averaged ITZ properties from those of the bulk paste are very different for the different ITZ parameters. The proportional effect is great for the content of unhydrated cement, which we find to be highly deficient in the innermost 10 μm wide strip, the mean amount being less than one-third that of that found in bulk paste. The proportional excess in pore space and calcium hydroxide contents within the ITZ are much more modest.

7. Specifically, we find that pore contents of 30% and higher found by others in the inner portions of the ITZs of various concretes have no counterparts in our measurements.

8. Our data indicate that the locally variable or 'patchy' character of the hardened cement paste within any particular shell of the ITZ is a reflection of the equally patchy character of the bulk hardened cement paste.

9. Concretes mixed only briefly develop hardened paste of much higher contents of detectable pores in both bulk and ITZ regions than well-mixed concretes of the same kind.

10. Models which invoke percolation as a necessary consequence of geometrical overlap between ITZs of adjacent grains appear to be misleading. If our results are correct, percolation could occur only when patches of locally high porosity within a particular ITZ overlap patches of similar high porosity in the ITZ of an adjacent grain. Such occasional local conjunctions are unlikely to form a continuous percolative chain across the depth of a concrete member, even for concretes whose statistical ITZ zones overlap appreciably.

11. In view of the patchy character of properties within the ITZ, the expectations for mechanical effects attributable to differing overlaps of ITZs in different concretes may also be overestimated.

ACKNOWLEDGMENTS

This paper is a contribution from the Purdue University unit of the Science and Technology Center for Advanced Cement-Based Materials. We thank the U. S. National Science Foundation for its continued support of this activity.

We gratefully acknowledge technical assistance provided by Janet Lovell and many discussions and assistance with computer problems provided by Jan Olek.

REFERENCES

1. J. Farran, Rev. Mater.des Construction, 490-492, (1956).

2. J. P. Ollivier, J. C. Maso, and B. Bourdette, Adv. Cement-Based Matls. 2 30-38 (1995).

3. A. U. Nilsen and P. J. M. Monteiro, Cem. Concr. Res. 23 147- 151 (1993).

4. K. L. Scrivener and K. M. Nemati, Cem. Concr. Res. 26 35-40 (1996).

5. S. Diamond, S. Mindess, and J. Lovell, in Liaisons Pates des Ciment Materiaux Assocoes, Proc. Colloque RILEM, Toulouse, C42-46 (1982).

6. E. J. Garboczi and D. P. Bentz, Adv. Cement Based Matls. 6 99-108 (1997).

7. K. L. Scrivener and P. L. Pratt, in Proc. 8th Intl. Cong. Chem. Cement, Rio, III 466-471 (1988).

8. Y. Wang and S. Diamond, in Mater. Res. Soc. Proc. 370 23-32 (1995).

9. H. P. Klug and L. E. Alexander, X-Ray Diffraction Procedures, Wiley and Sons, Inc. (1954), p. 271.

10. K. L. Scrivener, A. Bentur, and P. L. Pratt, Adv. In Cement Res. 1 230-237 (1988).

11. K. L. Scrivener, A. K. Crumbie, and P. L. Pratt, Mater. Res. Soc. Proc. 114 87-88 (1988).

12. K. L. Scrivener and E. M. Gartner, Mater.Res. Sco. Proc. 114 77-86 (1988).

13. P. Rangaraju, Ph.D. thesis, School of Civil Engineering, Purdue University (1997).

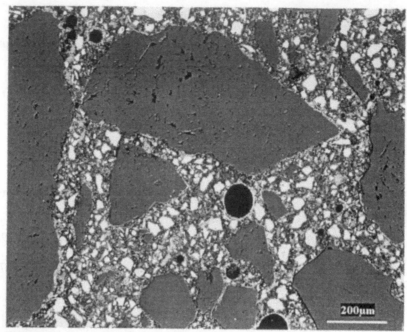

Figure 1.   Backscatter SEM micrograph from a 3 day-old brief-mixed w:c 0.50 concrete containing limestone sand and coarse aggregate.

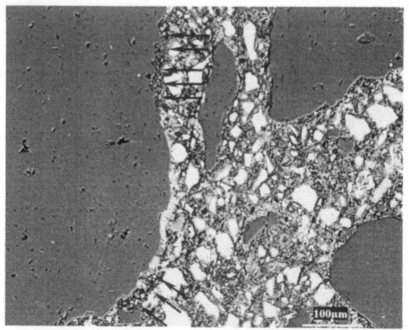

Figure 2   A portion of the left hand part of Figure 1 taken at higher magnification to show variations along the aggregate-paste interface.

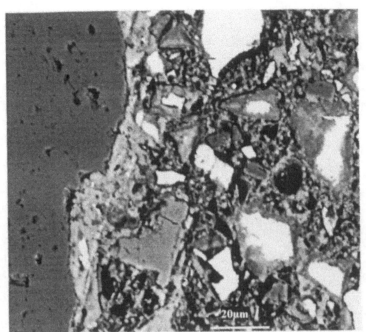

Figure 3.   A still higher magnification view of the central portion of Figure 2, taken to show details of the aggregate-cement paste interfacial region and various features of the structure of the hardened cement paste.

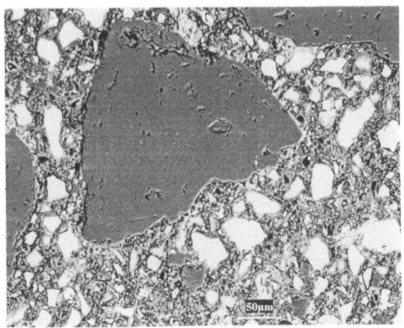

Figure 4.   Area of paste surrounding the triangular sand grain in the left-of-center portion of Figure 1.

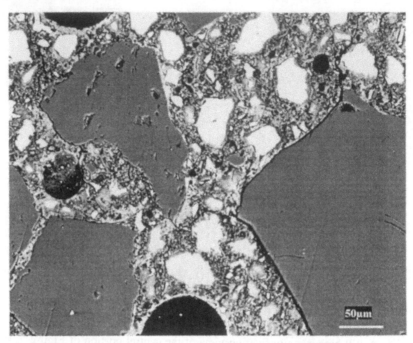

Figure 5.   Higher magnification view of the lower right hand portion of Figure 1, taken to show the features of a bond crack.

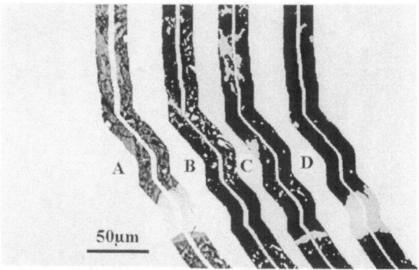

Figure 6.   Two adjacent sampling units taken from the image analysis for a 3 day-old, brief-mixed w:c 0.50 concrete containing quartzite sand and coarse aggregate. The sampling units are respectively 40 - 50 μm and 50 - 60 μm from the sand grain. The 'A' units are as originally secured; the 'B' units show only pore pixels; the 'C' units show only calcium hydroxide pixels, and the 'D' units show only unhydrated cement grain pixels.

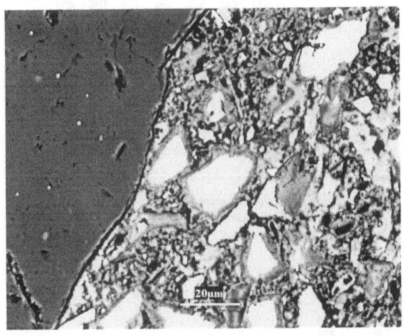

Figure 7. High magnification view of the paste region adjacent to a sand grain in a 3 day-old well mixed w:c 0.50 concrete containing quartzite sand and coarse aggregate. Comparison is invited to the paste microstructure shown in Figure 3 for a brief-mixed concrete.

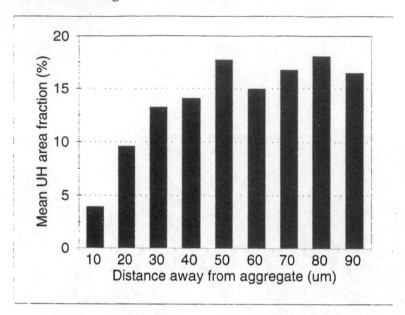

Figure 8.   Mean area% of unhydrated cement in successive 10 μm-wide strips around sand grains in the well-mixed concrete of Figure 7.

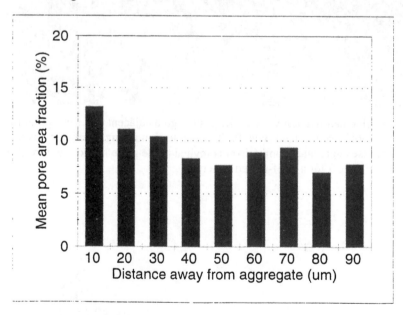

Figure 9.   Mean area% of pores detected in successive 10 μm-wide strips around sand grains in the well-mixed concrete of Figure 7.

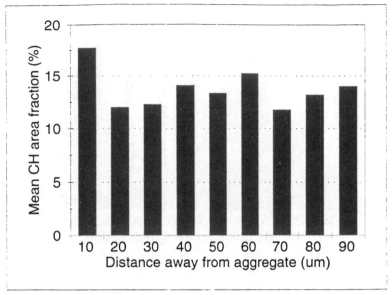

Figure 10. Mean area% of calcium hydroxide detected in successive 10 μm-wide strips around sand grains in the well-mixed concrete of Figure 7.

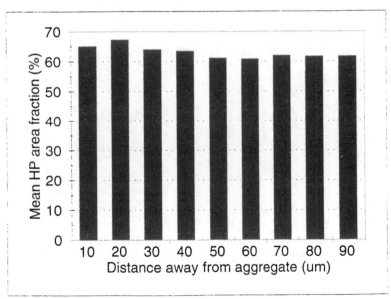

Figure 11. Mean area% attributed to 'hydrated cement' in successive 10 μm-wide strips around sand grains in the well-mixed concrete of Figure 7.

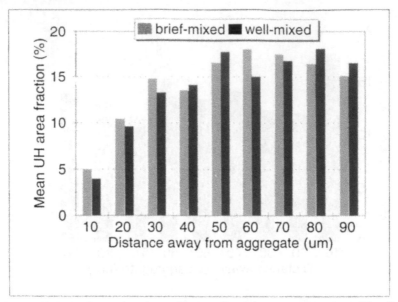

Figure 12   Split-bar chart showing effects of mixing on mean area% of calcium hydroxide in successive 10 μm-wide strips around sand grains.

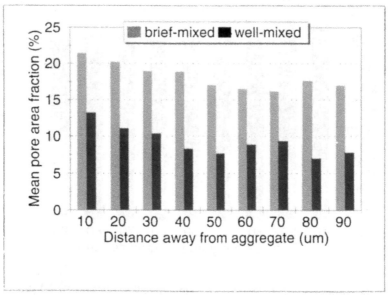

Figure 13.   Split-bar chart showing effects of mixing on mean area% of pores in successive 10 μm-wide strips around sand grains.

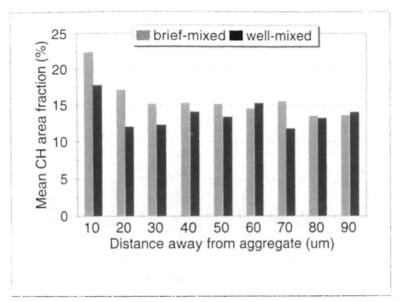

Figure 14. Split-bar chart showing effects of mixing on mean area% of calcium hydroxide in successive 10 μm-wide strips around sand grains.

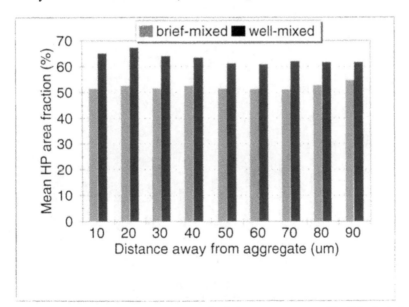

Figure 15. Split-bar chart showing effects of mixing on mean area% of hydrated cement' in successive 10 μm-wide strips around sand grains.

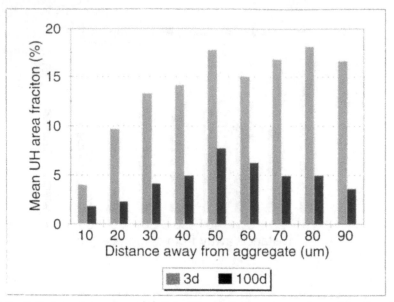

Figure 16. Split-bar chart showing the effects of age on mean area% of unhydrated cement in successive 10 μm-wide strips around sand grains.

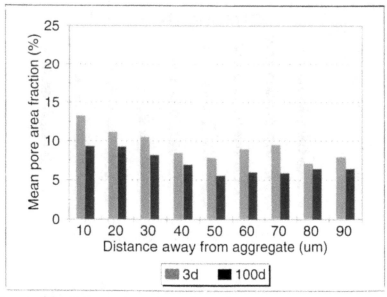

Figure 17. Split-bar chart showing the effects of age on mean area% of pores in successive 10 μm-wide strips around sand grains.

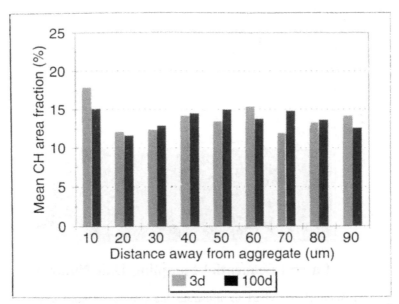

Figure 18. Split-bar chart showing the effects of age on mean area% of calcium hydroxide in successive 10 μm-wide strips around sand grains.

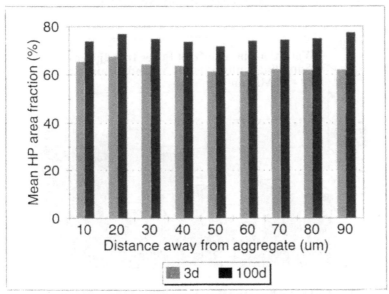

Figure 19. Split-bar chart showing the effects of age on mean area% of 'hydrated cement' in successive 10 μm-wide strips around sand grains.

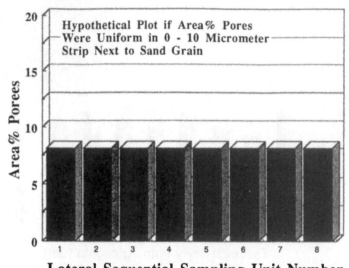

Figure 20. Hypothetical bar chart to illustrate the area% of pore space in adjacent sampling units constituting the 0-10μm strip around a particular sand grain in the concrete of Figure 1. The particular sand grain had 8 adjacent sampling units in the strip, and the mean porosity was 8.13 area%.

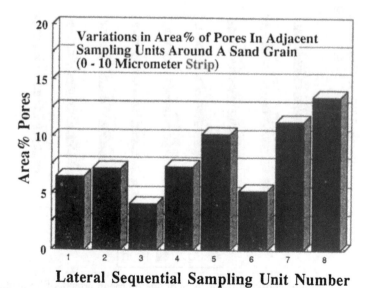

Figure 21. Area% pores found in the eight adjacent sampling units of the 0-10 μm strip around a sand grain whose mean pore content was 8.13 area%.

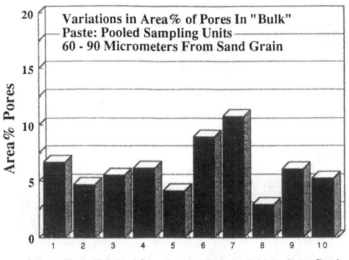

Sampling Unit Numbers, Pooled 6th-9th Strips

Figure 22. Area% pores found in the 10 non-adjacent sampling units of the bulk paste near the sand grain of Figure 21.

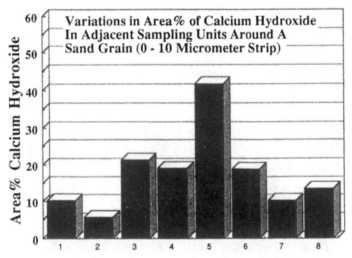

Lateral Sequential Sampling Unit Number

Figure 23. Area% calcium hydroxide in the eight adjacent sampling units of the 0-10 μm strip of Figure 21.

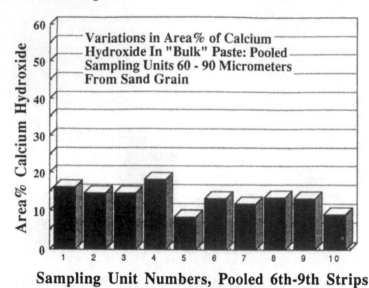

Figure 24. Area% calcium hydroxide found in the non-adjacent sampling units of the bulk paste of Figure 22.

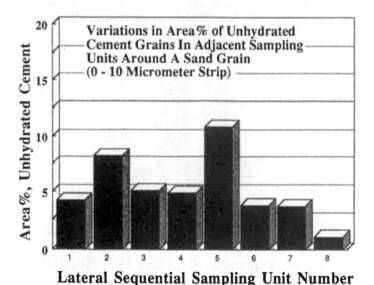

Figure 25. Area% unhydrated cement in the eight adjacent sampling units of the 0-10 μm strip of Figure 21.

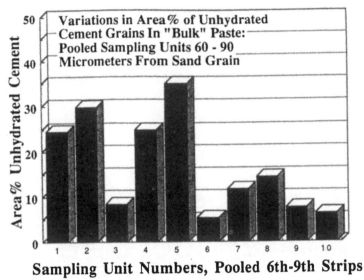

Figure 26. Area% unhydrated cement found in the non-adjacent sampling units of the bulk paste of Figure 22.

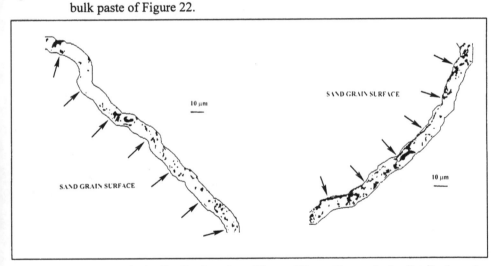

Figure 27a. Pore pixel distribution in sampling unit TV3e8-1, within the 0 - 10 μm strip around a sand grain in a 3 day-old well-mixed limestone-bearing concrete. No significant difference is noted between the 0 - 5 μm and the 5 - 10 μm layer portions within this unit.

Figure 27b. Pore pixel distribution in sampling unit TV3e6-1, one sampling unit away from that of Figure 27a. Excess pore pixels in the 0 - 5 μm layer portion are concentrated at the actual interface with the sand grain. Note: location of the interface may be inexact by the width of a pixel (0.5 μm)

PART TWO
MODELING, CHARACTERIZATION AND QUANTIFICATION OF ITZ STRUCTURE AND PROPERTIES

2 CONCRETE: A MULTI-SCALE INTERACTIVE COMPOSITE

E.J. GARBOCZI and D.P. BENTZ
Building and Fire Research Laboratory, National Institute of Standards and Technology, Gaithersburg, MD, USA

Abstract

Methods have been developed for predicting the overall diffusion coefficient of ions in concrete using a multi-scale interactive analysis, ignoring any chloride binding interactions. The analysis makes use of microstructure models of mortar/concrete and cement paste at the scales of millimeters and micrometers, respectively. When comparing experimental data to model results, one must be careful to consider the differing w/c ratio and degree of hydration in the interfacial transition zone (ITZ) and bulk paste regions, relative to a cement paste specimen containing no aggregates. The amount of aggregates influences the amount and properties of these two cement paste phases, leading to the designation of concrete as an *interactive composite*. The multi-scale analysis presented in this paper can quantitatively take this interaction of aggregates and cement paste into account when modelling the properties of concrete in terms of its microstructure. The analysis predicts that the contrast in diffusivity between the ITZ and the bulk cement paste is a function of the degree of hydration.

Keywords: Concrete, diffusivity, interfacial transition zone, microstructure, modelling, multi-scale

1 Introduction

One example of the complexity of the microstructure of concrete is found in the nature of the interfacial transition zones (ITZ) existing between aggregate and cement paste [1]. Due to an inefficient packing of cement particles near the aggregate, the ITZ regions are generally more porous and contain less unhydrated cement than the surrounding paste [2, 3]. This gradient of microstructure modifies the mechanical and transport properties of the concrete, so that the concrete must be considered as (at least) a three-phase material and not simply as a composite of aggregates in a matrix [4, 5, 6, 7].

The Interfacial Transition Zone in Cementitious Composites, edited by A. Katz, A. Bentur, M. Alexander and G. Arliguie. Published in 1998 by E & FN Spon, 11 New Fetter Lane, London EC4P 4EE, UK, ISBN: 0 419 24310 0

The large range of length scales necessary for modelling concrete require a multi-scale approach, where separate models have been developed for the millimeter (mortar/concrete) and micrometer (cement paste) scales [8]. By integrating these models, a complete calculation of the chloride diffusivity of a concrete may be obtained [9, 10]. This paper reviews this multi-scale analysis and presents evidence, validated by experimental data, that the contrast in properties between ITZ and bulk cement paste is a function of the degree of hydration and goes through a maximum as the cement hydrates. Needed refinements in the multi-scale analysis, as shown by experimental evidence, are also discussed.

2   Microstructural Modelling and Computational Techniques

2.1   Concrete Microstructure Model

For modelling concrete, a computational volume or box, of side dimension anywhere between 10 and 50 mm, is filled with spherical aggregates, each surrounded by a constant thickness ITZ [10, 11, 12], as shown in the left side of Figure 1. This model is a continuum model, with each aggregate completely characterized by the coordinates of its center and the value of its radius. Once a microstructure has been created, the volume fractions of ITZ and bulk paste (paste outside of the ITZ regions) and the degree of connectivity of the ITZ regions are determined [9, 10, 11] numerically or analytically. The connectivity of these ITZ regions across a 3-D concrete microstructure has recently been verified experimentally [13].

2.2   Cement Paste Microstructure Model

A cement hydration model is used to simulate and analyze the microstructure and properties of a single ITZ [14]. The cement powder to be modelled is represented by non-overlapping digitized spheres following the particle size distribution (PSD) measured on actual cement samples, with each pixel element representing 1 μm^3 in volume. A single flat plate aggregate is placed in the center of the microstructure before placing any of the cement particles (see lower part of the right side of Figure 1). The dimensions of the computational box are adjusted so that the appropriate volume ratio of ITZ to bulk cement paste is obtained, matching that previously determined using the concrete microstructure model.

After initial particle placement, microstructure development due to the hydration reactions between cement (tricalcium silicate) and water are modelled [14]. At any degree of hydration, the porosity present as a function of distance from the aggregate surface can be determined. Initially, after particle placement, the ITZ region contains a higher w/c ratio (more porosity) than the bulk paste due to the inefficient packing of the cement particles. During hydration, the porosity is reduced throughout, but it still remains higher in the ITZ regions. Thus, these regions will have a higher diffusivity than the bulk paste regions. The relative diffusivity (D/D_0) as a function of distance from the aggregate surface, x, can be

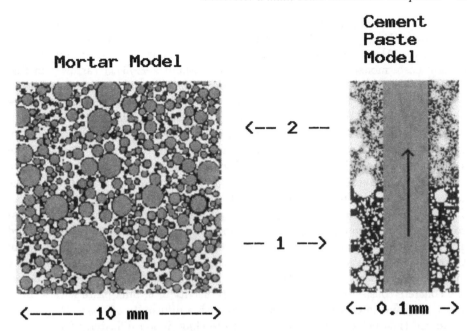

Figure 1: Linkages between microstructure models for mortar/concrete and cement paste in the multi-scale analysis. In the mortar model, inclusions are grey, ITZ regions are black, and bulk paste is white. In the cement paste model, cement particles are white, water-filled porosity is black, and the flat rectangular aggregate is grey. In the hydrated image (upper right), calcium hydroxide is dark grey and calcium silicate hydrate gel is light grey. The arrows indicate the flow of information between the models, and are explained in the text. The dimensions given are typical for each kind of model.

estimated using the equation [15]:

$$\frac{D}{D_0}(x) = 0.001 + 0.07 \cdot \phi(x)^2 + 1.8 \cdot H(\phi(x) - 0.18) \cdot (\phi(x) - 0.18)^2 \quad (1)$$

where relative diffusivity is defined as the ratio of the diffusivity of ions in the material of interest relative to their value in bulk water, $\phi(x)$ is the porosity fraction at a distance x, and H is the Heaviside function having a value of 1 when $\phi > 0.18$ and a value of 0 otherwise.

2.3 Multi-Scale Simulation Procedure

The multi-scale analysis is illustrated for a mortar in Figure 1 (the arrows in Fig. 1 show the flow of information between the two microstructure models and their parts). Initially, the median particle diameter of the cement PSD is used to establish the ITZ thickness, t_{ITZ} [12]. Aggregate particles following the aggregate

PSD are placed into the mortar volume. The volume fractions of interfacial transition zone (V_{ITZ}) and bulk (V_{bulk}) paste for this choice of aggregate PSD and t_{ITZ} [9, 10, 11] are then determined. The computational box dimensions for the cement paste model are chosen to match this ratio of V_{ITZ}/V_{bulk} (arrow 1 in Fig. 1). Cement particles are placed into this computational volume to achieve the desired w/c ratio, and the hydration model is executed to achieve the chosen degree of hydration (vertical arrow). It is by this step that the redistribution of cement between ITZ and bulk cement paste is approximately computed. The porosity of the cement paste is then measured as a function of distance from the aggregate surface and converted to relative diffusivity values using Eqn. 1. These values are averaged in two subsets, those lying within t_{ITZ} of the aggregate and those in the bulk paste, in order to give the values of D_{ITZ}/D_0 and D_{bulk}/D_0. The ratio of these two diffusivities, D_{ITZ}/D_{bulk}, is then used as an input back into the original concrete model (arrow 2 in Fig. 1). Of course, using a different value of t_{ITZ} would give a different value of D_{ITZ}/D_{bulk} [16]. Using the median particle diameter of the cement is probably the most realistic measure to use, however. The diffusivity of the overall concrete system, D_{conc}/D_{bulk}, is then computed [9, 10] At this point, the concrete consists of aggregates with a diffusivity of 0, bulk paste with a diffusivity of 1, and interfacial transition zones with a diffusivity of D_{ITZ}/D_{bulk}. This value can then be converted into an absolute chloride ion diffusivity for the concrete, D_{conc}, by multiplying it by D_{bulk}/D_0 as determined from the cement paste microstructure model and by D_0, the diffusion coefficient of chloride ions in bulk water [9, 17].

3   Results

3.1   Overall concrete diffusivity

Because of the local rearrangement of the cement particles in the vicinity of each aggregate, the ITZ w/c ratio will be significantly higher than that of the bulk cement paste in a concrete. Since the overall w/c ratio of a concrete mixture is determined at mix time, the bulk w/c ratio will therefore be less than this nominal value. To illustrate this effect, Figure 2 provides a plot of the initial distribution of anhydrous cement and water in a typical concrete with an ITZ thickness of 20 μm. The w/c ratio in portions of the ITZ region is seen to be over twice that of the nominal value of 0.45 and the bulk paste is seen to have a w/c ratio on the order of 0.4, quite similar to experimental results by Scrivener and Pratt in Ref. [1]. As the surface area of the aggregates increases, there will be more ITZ paste in a concrete, and a greater decrease in the bulk w/c ratio relative to the nominal value. Because the surface area to volume ratio of an aggregate increases with decreasing size, this means that there should be a larger ITZ effect on properties in mortars than in concrete.

The overall diffusion coefficient of a concrete is determined by the competition between the value of D_{ITZ}/D_{bulk}, the value of D_{bulk}, and the zero diffusivity of the aggregates. In most concretes, the zero diffusivity of the aggregates and the

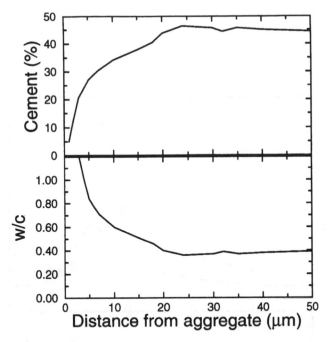

Figure 2: Initial distribution of anhydrous cement and w/c ratio in the ITZ for a concrete (overall w/c=0.45, 67.5% aggregate volume fraction.

value of D_{bulk} wins, so that the diffusivity of concrete is decreased by adding more aggregate, and has a lower diffusion coefficient than a cement paste of the same nominal w/c ratio. If the diffusivity of the ITZ regions were high enough, adding more aggregates to the concrete could actually increase the chloride ion diffusion coefficient above that of the equivalent cement paste [5]. This may be the case in some fine-sand mortars [18]. When considering fluid permeability, the usual case is for the concrete to have a much higher permeability than cement paste with the same nominal w/c ratio [19]. It is likely that the ITZ to bulk cement paste contrast is much higher in this case, so that adding aggregates would probably increase the permeability of the concrete.

3.2 D_{ITZ}/D_{bulk} vs. degree of hydration

The multi-scale model computes the value of D_{ITZ}/D_{bulk} for different degrees of hydration α. This parameter has sometimes been discussed as if it were a single number and not a function of α. However, eq. (1), which gives the diffusivity of cement paste as a function of porosity, is a non-linear function. Also, the porosities in the ITZ and bulk regions are simple, but different, linear functions of α. Therefore, the ratio D_{ITZ}/D_{bulk} is also a function of α.

Figure 3 shows model results for this ratio as a function of α for two different

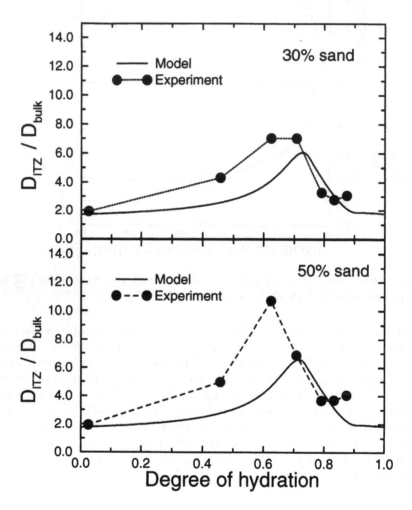

Figure 3: The computed value of D_{ITZ}/D_{bulk} as function of the degree of hydration for two mortars with the same cement and the same aggregate PSD, but with 30% and 50% sand volume fraction.

mortars, one with 30% and one with 50% sand by volume. The peak in the model results comes at about $\alpha = 0.70$. The experimental data points are from recent work on the electrical conductivity of mortars as a function of sand content and ITZ properties [20] (recall that electrical conductivity and ionic diffusivity are mathematically equivalent in this case). Reasonably good agreement is seen for the 30% sand mortar, with a greater difference between model and experiment for the 50% sand content mortar. In particular, the height of the experimental curve is not reached by the model prediction for the 50% sand mortar. Also, the experimental curves tend to peak slightly earlier in α for both mortars. At late hydration, the agreement between model and experiment is much improved. For durability studies, the predictions of the model will then be more reliable, since for service lives in years, higher degrees of hydration undoubtedly have occurred. Further detailed comparison between experiment and model data is found in Ref. [20].

4   Discussion and Summary

The agreement between model and experiment seen in Fig. 3, especially after substantial hydration, is encouraging. The discrepancies show that the model needs some refinement, however. In particular, eq. (1) probably needs to be looked at again, in light of improvements to the cement paste microstructure model made since it was developed [14]. Also, the way the parameter D_{ITZ}/D_{bulk} was extracted from experiment was by using a differential effective medium theory [10, 20]. This effective medium theory can also most likely be improved, which would affect the values of this parameter extracted from experiment . Finally, the distribution of hydration products between ITZ and bulk cement paste is probably not handled quite correctly at intermediate degrees of hydration by the cement paste microstructure model, judging by the disagreement shown in Fig. 3. The model is flexible in this area, and can probably be calibrated to predict this distribution more accurately.

Multi-scale microstructure models have been applied to simulating the development of ITZ microstructure in concrete and computing its effects on diffusivity. The model provides quantitative insights into the increased w/c ratio and porosity of the ITZ regions and the concurrent reduction in bulk paste w/c ratio and porosity. Details of the model can be systematically improved, which may improve the agreement with experiment.

REFERENCES

[1] Interfaces in Cementitious Composites, (ed. J.C. Maso), E&FN Spon, London, 1992; 1996.

[2] Scrivener, K.L., Bentur, A., and Pratt, P.L., "Quantitative Characterization of the Transition Zone in High Strength Concrete", *Advances in Cement Research*, **1** (4), 230-237, 1988.

[3] Bentz, D.P., Stutzman, P.E., and Garboczi, E.J., "Experimental and Simulation Studies of the Interfacial Zone in Concrete," *Cement and Concrete Research*, **22** (5), 891-902, 1992.

[4] Nilsen, A.U., and Monteiro, P.J.M., "Concrete: A Three Phase Material," *Cement and Concrete Research*, **23**, 147-151, 1993.

[5] Garboczi, E.J., Schwartz, L.M., and Bentz, D.P., "Modelling the Influence of the Interfacial Zone on the Conductivity and Diffusivity of Concrete," *Journal of Advanced Cement-Based Materials*, **2**, 169-181, 1995.

[6] Schwartz, L.M., Garboczi, E.J., and Bentz, D.P., "Interfacial Transport in Porous Media: Application to D.C. Electrical Conductivity of Mortars," *Journal of Applied Physics*, **78** (10), 5898-5908, 1995.

[7] Ollivier, J.P., Maso, J.C., and Bourdette, B., "Interfacial Transition Zone in Concrete," *Journal of Advanced Cement-Based Materials*, **2**, 30-38, 1995.

[8] Bentz, D.P., Schlangen, E., and Garboczi, E.J., "Computer Simulation of Interfacial Zone Microstructure and Its Effect on the Properties of Cement-Based Composites," in Materials Science of Concrete IV, Eds. J.P. Skalny and S. Mindess (American Ceramic Society, Westerville, OH, 1995) 155-200.

[9] Bentz, D.P., Garboczi, E.J., and Lagergren, E.S., "Multi-Scale Microstructural Modelling of Concrete Diffusivity: Identification of Significant Variables," *Cement, Concrete, and Aggregates*, in press (1997). Also available at http://ciks.cbt.nist.gov/garboczi/, Chapter 7.

[10] E.J. Garboczi and D.P. Bentz, J. of Adv. Cem-Based Mater., in press (1997).

[11] Winslow, D.N., Cohen, M.D., Bentz, D.P., Snyder, K.A., and Garboczi, E.J., "Percolation and Pore Structure in Mortars and Concrete," *Cement and Concrete Research*, **24**, 25-37, 1994.

[12] Bentz, D.P., Stutzman, P.E., and Garboczi, E.J., "Computer Modelling of the Interfacial Zone in Concrete," pp. 107-116 in Ref. 1.

[13] Scrivener, K.L., and Nemati, K.M., "The Percolation of Pore Space in the Cement Paste/ Aggregate Interfacial Zone of Concrete," *Cement and Concrete Research*, **26** (1), 35-40, 1996.

[14] Bentz, D.P. (1996). J. Amer. Ceram. Soc. 80, 3-21 (1997).

[15] Garboczi, E.J., and Bentz, D.P., "Computer Simulation of the Diffusivity of Cement-Based Materials," *Journal of Materials Science*, **27**, 2083-2092, 1992.

[16] E.J. Garboczi and D.P. Bentz, Analytical formulas for interfacial transition zone properties, J. Adv. Cem.-Based Mater., in press (1997).

[17] Mills, R., and Lobo, V.M.M., Self-Diffusion in Electrolyte Solutions (Elsevier, Amsterdam, 1989) p. 317.

[18] Halmickova, P., Detwiler, R.J., Bentz, D.P., and Garboczi, E.J. (1995) Water permeability and chloride diffusion in portland cement mortars: Relationship to sand content and critical pore diameter. Cem. and Conc. Res., Vol. 25,pp. 790-802.

[19] J.F. Young, in ACI SP108-1, *Permeability of Concrete* (1988), ACI, Detroit.

[20] Shane, J., Mason, T.O., Bentz, D.P., and Garboczi, E.J. (1997) Experimental and theoretical study of mortar conductivity, in preparation.

3 FRACTURE OF THE BOND BETWEEN AGGREGATE AND MATRIX: AN EXPERIMENTAL AND NUMERICAL STUDY

A. VERVUURT[1] and J.G.M. VAN MIER
Delft University of Technology, The Netherlands
[1]Current working address: TNO Building and Construction Research

Abstract
Experiments have been carried out in order to determine the input parameters for the Delft Lattice Model. Emphasis is placed on the fracture properties of the interfacial transition zone between matrix and aggregate particles. Both the experiments and the simulations presented in this paper have been performed at the Stevin Laboratory of Delft University of Technology. In the experiments a single aggregate particle is positioned at a prescribed position of a notched specimen. Interface fracture is obtained by a splitting load. To vary the interface characteristics, either a granite or a sandstone aggregate particle is used. The results are used to determine the interface properties in the Delft Lattice Model. In the two dimensional model concrete is schematised as a three phase material by means of brittle breaking beam elements. Fracture is simulated by removing an element as soon as the strength of the element is reached. The results show that the stiffness of the beam elements hardly affects the fracture process, whereas the failure mechanism is dominated by the strength of the beams.

1 Introduction

For heterogeneous materials like concrete it is widely accepted that the interfacial transition zone is the most critical region for crack initiation. When studying the interfacial transition zone in concrete it is essential that an unambiguous definition of this zone is given. In this respect it is noted that in this paper the mechanical properties of the interfacial transition zone between the (largest) aggregate particles and matrix (consisting of the hardened cement paste and the finest sand particles) are studied. From this definition it may be obvious that the research focuses on the particle level, also called meso level. Both at the meso level and the micro level interfaces are formed by a combination of physical, chemical and mechanical bonding [1]. At the meso level mechanical bonding seems to be dominating, whereas on the micro level physical and chemical bond is more pronounced [2]. At the macro level, on the other hand, concrete can be treated as a continuum material and no interface can be distinguished within the material. To describe the material behaviour at the latter level properly, a non linear

The Interfacial Transition Zone in Cementitious Composites, edited by A. Katz, A. Bentur,
M. Alexander and G. Arliguie. Published in 1998 by E & FN Spon, 11 New Fetter Lane,
London EC4P 4EE, UK, ISBN: 0 419 24310 0

constitutive material behaviour has to be assumed. In this paper it will be shown that macro-scopic non-linearities can be simulated at the meso-level by brittle behaviour of the individual composites. A difficulty in this approach is, however, that the properties of the modelled phases have to be known. For the aggregate and matrix phase this behaviour can be deter-mined rather straightforward, but for the interfacial transition this is not quite unequivocal. An often used method to study the mechanical properties of the interfacial transition zone is to isolate the aggregates by adopting only a few particles, which are positioned carefully at a prescribed position in a specimen [3], or by casting cement paste to a cored aggregate [4], [5]. From the approaches followed by several researchers it appeared, however, that there is still quite some disagreement in the exact role of the interfacial transition zone on the fracture be-haviour of concrete. Moreover, it appeared that the conditions of preparing, curing and testing the specimens is of major importance when evaluating the results. In the experiments and simulations presented in this paper careful attention was paid to consistency in preparing the specimens and carrying out the tests. In the simulations it was tried to model the tests as accu-rate as possible. The fracture properties required in the numerical model have been determined by carefully observing the fracture processes in the experiments with an optical microscope.

2 Outline of the program

2.1 Experimental program

The interfacial transition zone between the matrix and the aggregate particles in concrete is studied for two types of matrix material. Hardened cement paste mixtures, were tested along with mortars with a maximum aggregate size of the sand of 0.25 mm. For all specimens a white cement was chosen for maximum contrast with the cracks when using the long distance (optical) microscope that was used for monitoring crack growth. The microscope will be dis-cussed in detail further on in this paragraph. Three different water-cement ratios were tested for each material. Since the results of the specimens containing the different matrix materials did not differ substantially, no further attention will be paid to this aspect in this paper, and only the most significant results will be presented. A full overview on the test results, how-ever, is given in [6].

For studying the bond properties between the matrix and the aggregate, the matrix was cast around a selected aggregate material. The results concerning two types of aggregates are presented. A dense material was studied along with a porous kind of sandstone resembling a lightweight type of aggregate. For this purpose, aggregates were cored from blocks of Ben-theimer sandstone from Germany. Opposite to the porous sandstone, Polar White granite from Brazil was used as a dense aggregate material. The type of granite was selected because of its white colour. Specifications and a full description of the results concerning these experiments is reported in [6].

Several types of specimens were tested. Next to varying the type of matrix and aggregate material, the aggregate position, the number of aggregates and the particle shape were varied. In this paper, however, all experiments presented contained a single (cylindrical) aggregate particle at a fixed position in the specimen. The position of the particle with respect to the notch is illustrated in Fig. 1a. For reference each matrix material was also tested without any aggregates. The specimens with planar dimensions of 125x168 mm and a thickness of 15 mm, were notched at half width of the specimen, as indicated in Fig. 1a. The notch length was taken 30 mm, and was sawn over the full depth of the specimen. The chosen length of the notch mainly depended on the unrestricted view on the notch tip, as it was required by the op-tical microscope. In order to perform the experiments under displacement control, the Crack Mouth Opening Displacement (CMOD) was used as feedback signal for the servo controlled system. The CMOD was measured by means of two Linear Variable Differential Transducers (LVDTs) which were placed across the top of the notch (see Fig. 1b).

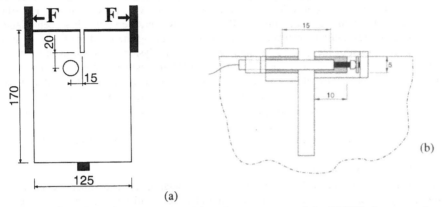

(a)

(b)

Fig. 1   Adopted specimen geometry (a) and arrangement of the LVDTs for measuring the Crack Mouth Opening Displacement (b).

The experimental set-up which was used for testing the specimens is shown in Fig. 2a. The test set-up roughly consists of two parts which are both illustrated in Fig. 2a. During testing the loading frame containing the specimen was centred under the loading device. The loading device was specially prepared in order to apply a perfectly horizontal load to the specimen. To circumvent the measurement of frictional forces, two load cells were positioned between the load cells and the specimen.

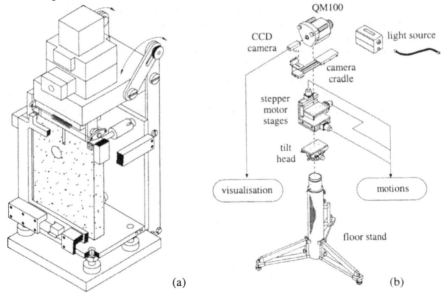

(a)   (b)

Fig. 2   Experimental set-up (a) consisting of a loading frame and loading device and (b) schematical view of the remote controlled long distance (optical) microscope.

To retrieve detailed information regarding the (micro-)cracks that develop at the surface of the specimen, real-time crack measurements have been performed with a high resolution long distance optical microscope. The microscope has proven to be very useful for explaining failure mechanisms in concrete. In early tests as for example presented in [7], the system was suitable for manual scanning only. Very labour intensive work was necessary in order to filter the crack from the monitored video images. Recently, however, the system has been adapted for digitising images and automatically storing and re-scanning the crack path [6]. A schematical representation of the QUESTAR Remote Measuring System is shown in Fig. 2b. The system consists of a QUESTAR microscope (QM100 MK-III) which is connected to an Ikegami (ICD-46E) black and white CCD Camera for retrieving the images. The microscope and the camera are fixed to a cradle and subsequently to stages which can be moved in three orthogonal directions. Each stage is equipped with a stepper motor (Compumotor A/AX57-51), with a range of approximately 50 mm and an accuracy of about 1 μm. During the test, the specimen is illuminated using two fibre optic arms, connected to a AIMS FB-150 light source.

2.2 The Delft Lattice Model

A numerical lattice model [8] has been adopted for simulating the experiments containing a single aggregate particle. In the model the material is schematised as a network of brittle breaking beam elements. To simulate failure of the material, a beam is instantaneously removed from the mesh as soon as the strength of that particular element is exceeded. For modelling the *single-particle-tests* as described in the previous paragraph, disorder is implemented in the mesh in two different ways. First a so called random lattice structure is adopted (see Fig. 3) and additionally, granular heterogeneity is assigned to the beams by means of an aggregate overlay on top of the lattice and by assigning different (fracture) properties to the three phases in the material (aggregate, matrix and the interface between the matrix and the aggregate, see Fig. 4)

In the random lattice adopted in this paper a regular square grid with cell size s is generated, after which in each cell of the grid a node is selected at random. Subsequently the (triangular) lattice configuration is determined by connecting always the three nodes which are closest to each other. The procedure for generating the lattice is illustrated in Fig. 3a. The randomness of the mesh is controlled by the parameter A, which limits the area in a cell where the node is selected. It may be obvious that for decreasing A, the randomness of the mesh decreases as well (as shown in the examples of Fig. 3b-d), and that in the limit case of $A=0$, a regular square lattice is obtained with crossing diagonals. The random lattice adopted in this paper is assigned with a cell size of $s=1$ mm and a randomness of $A=0.75 \cdot s$.

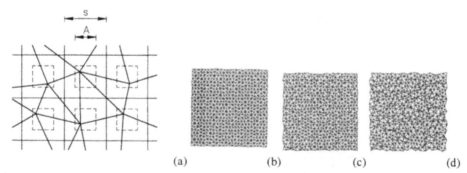

(a)　　　　　(b)　　　　　(c)　　　　　(d)

Fig. 3   Generation of a random lattice (a) and meshes with a randomness of A=0.2·s (b), A=0.5·s (b), and A=0.8·s (c).

In spite of the fact that a random lattice with constant material properties for the beams behaves heterogeneously by nature, an aggregate overlay was used to assign additional heterogeneity, as illustrated in Fig. 4. In practice this means that each individual phase (i.e. aggregates, matrix and bond) is assumed to behave heterogeneously as well. Note that this may also be achieved by assigning a statistical distribution of the beam properties within each phase of the three material phases. It has been shown that the heterogeneity of the material is characterised by the strength and stiffness ratios of the three material phases [6]. For the strength (f_t) these ratios are $f_{t;M}/f_{t;A}$ and $f_{t;M}/f_{t;B}$, whereas for the stiffness the ratios are $E_{t;M}/E_{t;A}$ and $E_{t;M}/E_{t;B}$, where the subscripts A, M and B stand for aggregate, matrix and bond between matrix and aggregate respectively.

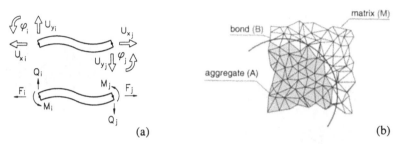

(a)   (b)

Fig. 4   Degrees of freedom and forces acting on the adopted beam elements (a) and (b) lattice projected on top of an aggregate particle for assigning the beam properties

The geometric and material properties of the beams are the height h, the thickness t, the Poisson's ratio v, and the Young modulus E, and depend only on the elastic behaviour of the lattice. The strength (f_t) of the beams, on the other hand, is directly related to the material's fracture behaviour. The determination of these properties can proceed in a rather straightforward manner as far as the beams in the matrix and the aggregate phase are concerned [6]. Because the interfacial transition zone can not be isolated from the material, it may be obvious that the determination of the interfacial properties is not as straightforward as for the matrix and the aggregates.

3 Results

In the presented simulations and experiments the influence of the altering strength and stiffness of the interfacial transition zone between matrix and aggregates, on the fracture behaviour of a specimen, is studied. In total over 60 simulations and 100 experiments were performed. In this paper, however, only the most significant results are presented. For a discussion on the complete series of simulations and experiments the reader is referred to [6]. With respect to the experiments, the simulations can be divided into two groups, namely those related to specimens containing a granite aggregate particle and those containing a sandstone aggregate particle. For each series of experiments three ratios $f_{t;M}/f_{t;B}$ and four ratios of $E_{t;M}/E_{t;B}$ were investigated for the lattice model. Next to varying the interfacial transition zone, a few additional parameters of the lattice model were studied, such as the mesh dependency and the position of the big aggregate particle. These results, however, are not discussed in this paper, but reported in [6].

The results for specimens containing a granite aggregate particle are shown in Fig. 5. In the figure the experimental results are given by means of the final crack pattern and a detail near the interfacial transition zone as obtained with the optical microscope. In Fig. 5c the numerically predicted crack pattern is shown. In both the experiments and the simulations it appeared that cracking is initiated at the notch. Due to the presence of the large aggregate particle, however, the crack arrests at the aggregate and cracking proceeds along the interfacial transition zone, as can clearly be seen in the microscope recording of Fig. 5b. In the simulations it appeared that the experimentally observed behaviour can be predicted quite well when the strength of the interfacial transition zone is substantially lower than the strength of the matrix (in the example of Fig. 5c the ratio $f_{t;M}/f_{t;B}$ was set to 4). Furthermore it appeared that the influence of the stiffness of the interfacial transition zone was of negligible influence when compared to the influence of the interfacial strength.

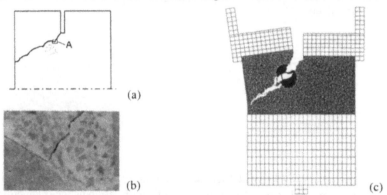

Fig. 5   Experimentally (a, b) and numerically (c) obtained crack pattern in a specimen containing a granite aggregate particle. In Figure (b) a detail of area A, as obtained with the long distance microscope, is shown.

When sandstone, instead of granite, is used as an aggregate material, the results as shown in Fig. 6 are obtained. The behaviour is substantially different from the results with a granite aggregate particle. Both the relative high interface strength and the low aggregate strength contribute to the fact that the crack penetrates the interfacial transition zone (see Fig. 6b).

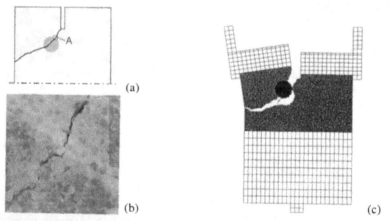

Fig. 6   Experimentally (a, b) and numerically (c) obtained crack pattern in a specimen containing a sandstone aggregate particle. In Figure (b) the area A of Figure (a) is shown enlarged as an image obtained with the long distance microscope.

In the simulations (Fig. 6c) the strength was again the controlling parameter for manipulating the results, and the interfacial stiffness was of minor importance. It appeared that the strength of the interfacial transition has to be chosen larger than the strength of the aggregate ($f_{t;B} > f_{t;M}$) in order to obtain results which compare favourable to the experimental observations.

The effect of the type of aggregate on the load-CMOD behaviour is given in Fig. 7. In the figures the results are plotted for the experiments as discussed above. Globally, the observed behaviour is comparable. In both experiments a diagram is obtained in which two peaks can be distinguished. From the microscope recordings it was concluded that the first peak in the diagram was observed as soon as the crack initiated at the notch of the specimen. After a drop of the load during propagation of the latter crack, a second ascending branch is observed which is explained from the fact that the crack arrests near the large aggregate particle. The second peak is observed as soon as the crack starts propagating towards the side of the specimen at which the aggregate is situated. Ahead of the second peak micro-cracks appear either in the interfacial transition zone (when a granite aggregate is adopted, see Fig. 6a) or in the aggregate particle (when a sandstone aggregate particle is used, see Fig. 6b). Due to the weak interfacial transition zone for the specimen containing a granite aggregate particle, a much lower second peak is observed than for specimens containing a sandstone aggregate particle.

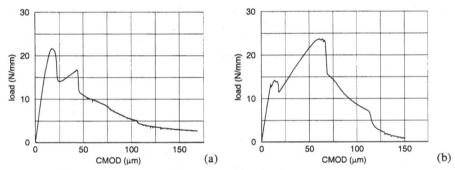

Fig. 7   Load-CMOD curves for a specimen containing a (a) granite and (b) sandstone aggregate particle.

4 Conclusions

A combined numerical and experimental approach is adopted for studying the mechanical properties of the interfacial transition zone between matrix and aggregate in heterogeneous materials like concrete. Splitting tensile experiments were performed in which a horizontal load was applied to a specimen containing a single, cylindrical aggregate particle, which was located at a prescribed position in the specimen. An optical microscope was adopted for monitoring crack growth in the interfacial region. In both the experiments and the simulations two types of aggregate with extreme interfacial properties were tested. A porous sandstone was used to imitate the behaviour of a light weight aggregate material whereas granite was used as a substitute for a river gravel aggregate. From the experiments it appeared that the interfacial zone for granite acts very weak, which is reflected by crack growth along the interfacial transition zone. For the sandstone aggregate particle the crack is more likely to penetrate the interface. In the simulations with the Delft Lattice Model this behaviour is dominated by the strength of the beams in the interfacial transition zone. The stiffness of these beam elements seems to have a minor effect on the fracture mechanism.

5 Acknowledgement

The authors are greatly indebted to Mr. A.S. Elgersma for his assistance in carrying out the experiments.

6 References

[1] Van Mier, J.G.M. (1997) *Fracture Processes in Concrete: assessment of material parameters for fracture models.* CRC Press, Inc.

[2] Zhang, M.H. and Gjørv, O.E. (1990) Micro-structure of the interfacial zone between lightweight aggregate and cement paste. *Cement and Concrete Research*, Vol. 20, pp. 610-618.

[3] Lee, K.M. and Buyukozturk, O. and Oumera, A. (1992) Fracture Analysis of Mortar-Aggregate Interfaces in Concrete. *Journal of Engineering Mechanics (ASCE)*, Vol. 18, No. 10, pp. 2031-2047.

[4] Alexander, K.M. (1959) Strength of the Cement-Aggregate Bond. *Journal of the American Concrete Institute*, Vol. 56, No. 11, pp. 377-390.

[5] Wang, J. and Maji, A.K. (1994) Experimental Studies and Modeling of the Concrete/Rock Interface. *ACI Special Publication on Interface Fracture and Bond*, Vol SP-156, pp. 45-68.

[6] Vervuurt, A. (1997) *Interface Fracture in Concrete.* PhD Thesis Delft University of Technology.

[7] Van Mier, J.G.M. (1991) Crack Face Bridging in Normal, High Strength and Lytag Concrete. In *Fracture Processes in Concrete Rock and Ceramics* (eds. Van Mier, J.G.M. and Rots, J.G. and Bakker, A.). Chapman & Hall E&FN Spon, pp. 27-40

[8] Schlangen, E and Van Mier, J.G.M. (1992) Experimental and numerical analysis of micro-mechanisms of fracture of cement-based composites. *Cement and Concrete Composites*, Vol. 14, pp. 105-118.

4 INTRODUCTION OF SYNERESIS IN CEMENT PASTE

M.R. DE ROOIJ and J.M.J.M. BIJEN
Department of Civil Engineering, Delft University of Technology, Delft,
The Netherlands
G. FRENS
Department of Chemistry, Delft University of Technology, Delft,
The Netherlands

Abstract
Although the microstructure of the interfacial zone is subject to lot of research, a complete and satisfactory explanation for the formation of the interfacial transition zone has not yet been given. All to easy the reason is said to be the wall-effect. However, based on earlier work at Delft University of Technology, it is thought that syneresis, a phenomenon known in colloid science, might be the answer. Syneresis leads to exudation of fluid of a colloid system.

'Active thin sections' are used in experiments to show the occurrence of water exudation in Portland cement paste. Furthermore, it is shown that blast furnace slag cement exudates substantially less water than ordinary Portland cements and thus leads to a smaller interfacial zone, as is known in literature. Also, it is shown that finely ground Portland cement expels more water than coarse.
Keywords: active thin sections, cement paste, formation, microstructure, syneresis.

1 Introduction

The importance of the interfacial transition zone and its influence on the behaviour of concrete has already been highlighted by RILEM with the organisation of an international conference in 1992, Toulouse, and a state-of-the-art report [1]. In the RILEM report, the interfacial zone is described from its microstructure and its properties all the way up to the (possible) influences on the behaviour of concrete as a whole. However, a fundamental explanation for the formation is not given.

In this paper the first part of an investigation into the formation mechanisms of the interfacial zone is presented. It is shown, that by introducing syneresis (a phenomenon

The Interfacial Transition Zone in Cementitious Composites, edited by A. Katz, A. Bentur,
M. Alexander and G. Arliguie. Published in 1998 by E & FN Spon, 11 New Fetter Lane,
London EC4P 4EE, UK, ISBN: 0 419 24310 0

known in the field of colloid science) in cement paste, strong evidence can be gathered to explain the formation of the interfacial zone.

2 Knowledge to date

The interfacial transition zone is part of the hydrated cement structure. Close to the aggregate surface the microstructure differs remarkably from the 'bulk' of hardened cement paste. The microstructural difference in ordinary Portland cement, for instance, can measure up to 50 μm and gradually dissolves going away from the aggregate. A schematical impression of the interfacial zone is given in figure 1.

Directly at the surface of the aggregate a thin layer of products is formed, typically only a micron or so in thickness. Barnes et al. [2] characterised this as a 'duplex film'. It consists of a rapidly precipitated layer of calcium hydroxide (CH) on an aggregate surface exposed to the mix water, and a thin single layer of short fibres C-S-H gel in an open parallel array on its paste-facing surface. Zimbelmann [3] only notes the network of fine ettringite crystals which have been deposited on the aggregate surface in the initial stage of hydration (i.e. up to an age of about ten hours).

Next, a 'contact layer' exists mainly of CH crystals, with the c-axis of their hexagonal unit cell perpendicular to the aggregate surface [4, 5]. According to Diamond [6], the contact layer is separated from the aggregate by the duplex film. The contact layer in his opinion is serving as an intermediate layer between the very thin duplex film and the sparsely populated region towards the bulk. The contact layer only develops after a day or two have elapsed.

On the side of the contact layer opposite to the aggregate, much solution-filled space remains in which special forms of hydration product develop. One often finds relatively large, hexagonal CH crystals, sometimes tens of microns across but only one or two microns thick, and clusters of ettringite needles [7]. The panel-shaped crystals and ettringite needles are anchored in a unit of hydrates of the neighbouring cement grains.

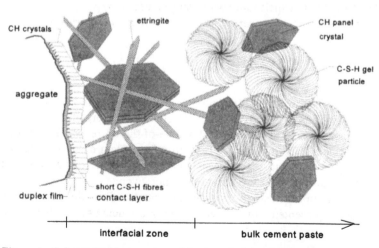

Figure 1. Schematical impression of the interfacial transition zone.

According to Zimbelmann the mainly needle- and rod-shaped ettringite hydrates rub against and adhere to the contact layer [4].

From the microstructure of the interfacial zone as it has been described here, typically, the high porosity, the large portlandite plates and the long ettringite needles are mentioned as characteristic in literature. Based upon especially these characteristics it has to be concluded, that the area in which the interfacial zone is formed, requires on that location a relatively high water/cement ratio.

It has to be pointed out here, that many theories on mechanisms of the formation of the interfacial zone remain in the area of remarks, rather than full paper explanations. Researchers that have investigated the microstructure and/or the composition of the interfacial zone, make a note on the possible cause of their observed phenomena, and mostly leave it there.

The simplest explanation comes from Zimbelmann [8]. He concludes, that immediately after introducing water to the solid particles in the microstructure, all the solid particles (aggregates, cement, etc.) are covered at once by a water film. To explain the size of the interfacial zone, Zimbelmann states that this water film possesses nearly a constant thickness at all particles of about 10 μm. This is hardly to believe.

Another explanation named localised bleeding was suggested by Scrivener and Pratt [9]. They report that the relative movement of sand and cement grains during mixing, and possibly setting of the sand grains before the cement paste sets, may lead to regions of low paste density at the interface. Bleed water accumulates beneath the larger aggregate particles, creating additional planes of weakness. This has indeed been confirmed in quantitative studies of the microstructure by Hoshino [10]. However, the interfacial transition zone is also present at the 'upper' side of aggregates. This means that the localised bleeding may help but is in fact not the primary mechanism for the formation of the interfacial zone.

The explanation currently most referred to is known as the 'wall-effect'. Mehta and Monteiro [11] used this term for the problem of particle-packing next to a 'wall' (i.e. aggregate surface). Certainly, the wall-effect will occur, but it can not be considered the entire explanation. Cement grains have a wide range of sizes ranging from 1 μm up to even 100 μm sometimes. The smaller particles fill up the holes between the bigger particles. Furthermore, simulations with particle packing do show a gradient in particle density at the interface, but only to an extent of 10 μm, which leaves the wall-effect unable to explain a thickness as large as 50 μm at the interfacial zone.

3 Syneresis

In recent years it has been realised more and more that the basics for the final properties of any concrete is laid during the first days and even the first hours of the hydration process. At Delft University of Technology work has been done on the effect of mineral admixtures (additions) on the microstructure of concrete [12, 13, 14]. It was concluded from that work, that colloid science is a key to explain the observed phenomena scientifically. In this field Yang et al. [15] introduced already the use of the DLVO theory (named after the four contributors: Derjaguin, Landau, Verwey and Overbeek), which consists of Van der Waals attraction and the electrostatic double

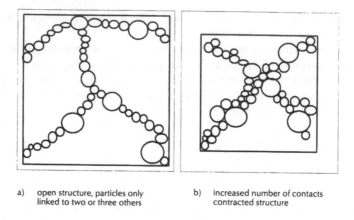

a) open structure, particles only   b) increased number of contacts
    linked to two or three others   contracted structure

Figure 2. Schematic presentation of the effect of syneresis.

layer repulsion, on cement systems. They have shown that neat pastes of ordinary Portland cement have an ionic concentration which is well above the critical coagulation concentration and therefore are subject to rapid coagulation. Where they left off, we continued to find an explanation for the formation of the interfacial transition zone.

When a colloidal sol becomes coagulated rapidly, the result is a very open structure in which most particles tend to be linked to only two or three other particles (see figure 2a). It contains a great deal of entrapped solvent (usually water) [16]. After the first rapid settling of a gel, the particles still retain some freedom of motion. A slow flocculation and reorganisation continues, against the strength of the gel, to increase the number of contacts per particle in order to decrease the free energy of the system. This process inevitably entails a contraction of the disperse phase (see figure 2b); the volume decreases and solvent is spontaneously pressed out. This phenomenon was first termed **syneresis** by Thomas Graham [17].

An example of syneresis can be found in the formation of a blood clot. When blood flows out of its vessel and comes into the open a complex system of reactions turns albumins into a network of fibrin fibres. Into this network platelets are trapped and they are mainly responsible for the contraction of the network to about half its original volume, meanwhile squeezing out the blood serum. This is a common example of syneresis.

Another example of syneresis can be found in the field of dairy-industry. Gels formed from milk by renneting or acidification under quiescent conditions may subsequently show syneresis, i.e. expel liquid (whey), because the gel (curd) contracts. Under quiescent conditions, a rennet-induced gel may lose two thirds of its volume, and up to 90% or even more, if external pressure is applied [18]. Often, syneresis is undesired, e.g. during storage of products like yoghurt, sour cream, cream cheese or quark. On the other hand, when making cheese from renneted or acidified milk, syneresis is an essential step.

Jefferis and Sheikh Bahai [19] have studied the syneresis in silicate-aluminate grouts. Initial tests on cylinders of gels showed that the gel tends to shrink in the direction of gravity, which led to speculation that syneresis is a gravity effect similar to bleeding in cement grouts or self-weight consolidations of clay. However, close examinations of the gel in the cylinders showed that the gel had contracted radially as well as axially.

For cheese making, attempts have been made to model the process of syneresis. Van Dijk [20] has studied horizontal slabs of renneted milk. The diameter of the cylindrical slabs was much larger than their thickness. This way, one-dimensional syneresis under constant conditions could be determined. Unfortunately, the measurements turned out to be to difficult. The numbers were too small and they varied with ongoing syneresis. In spite of these difficulties it was still attempted to calculate a rate of syneresis, but the only conclusion which would hold was, that under constant conditions, the rate of syneresis decreases as syneresis proceeds.

If syneresis occurs in cement paste, it leads to considerable shrinkage of the cement gel forcing water molecules out of the gel structure. This leads to a redistribution in the originally homogeneous mass of cast cement in water-rich and solids-rich areas. These conditions - local variation of the water/cement ratio - translate into variations in the cement hydration process which follow at filling the formwork and which eventually determine the structure of cement and concrete. The formation of the interfacial transition zone can thus be explained.

4 Experiments

To investigate the hypothesis of syneresis, in analogy with the experiments on slabs of renneted milk, an 'active thin section' of cement in which the cement is actually hydrating was examined (figure 3). A drop of cement paste is placed on a slide object holder and studied with optical microscopy. To prevent water from evaporating and to avoid interference by carbon dioxide from the atmosphere, the cement drop is sealed off with double-sided adhesive tape and a cover glass. A quadrangle of plastic foil is used as a spacer. When a drop of cement paste is placed in the hole (see figure 3) and the cover glass is pressed on to it, a two-dimensional syneresis system is created.

When the rim of the cement drop is continuously examined during the first hours of hydration, water can be seen to emerge. The area of the water that has flown from the cement drop is measured as a function of time using image analysis. A composite image for various times of hydration is presented in figure 4. The total area of exuded

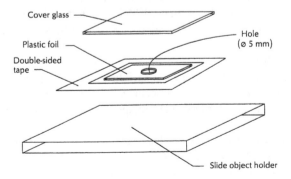

Figure 3. Schemetical build up of an active thin section.

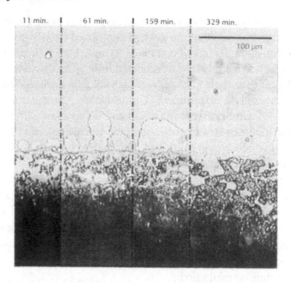

Figure 4. Composed image of exudating water while
cement retreats for 11, 61, 159 and 329 minutes after
first contact between cement and water.

water is related to the original cement drop area in order to arrive at a formed water
area percentage. Some initial results with various cements are shown in figure 5. The
water/cement ratio was 0.5 in all experiments.

It is known that blast furnace slag has a smaller interfacial zone, than Portland cements [12]. Our results show that the blast furnace slag cement has substantially less
water exudation than the Portland cements investigated. The cause of this difference in
exudation is not clear, although it is known that ground blast furnace slag cement
particles do have a different zeta-potential from Portland cement particles [21], which
does agree with the observed phenomena.

Furthermore, we see that the finely ground Portland cement CEM I 52.5 R expels
more water than Portland cement CEM I 32.5 R. This suggests that a larger interfacial
zone for the finer cement can be expected. The contrary has been predicted on the
basis of particle-packing at walls. However, in view of the colloid chemical theory for
syneresis we would have predicted that CEM I 52.5 R should expel more water than
CEM I 32.5 R. Since no literature is known to us, we have started experiments for
verification of this point.

Besides the expulsion of water, also the contraction of the cement gel itself is observed during the experiment. The total shrinkage should be due to two effects: the
chemical shrinkage which is known to occur during the hydration of cement [22] and
the colloidal syneresis. The observed shrinkage seems to exceed the chemical shrinkage that is known.

It is known that the interfacial zone is less extensive than in ordinary Portland cement
if additions are used like silica fume and fly ash particles [12, 13]. This is partly due to
the puzzolanic effect and the fact that silica fume and fly ash particles are very small,
which results in an even better particle-packing. However, they also influence the

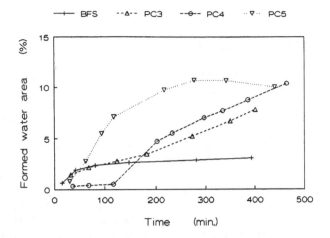

Figure 5. Formed water area for various cements. BSF = Blast Furnace Slag cement, PC3 = CEM I 32.5 R, PC4 = CEM I 42.5 R, PC5 = CEM I 52.5 R.

electrostatic potential of cement. Furthermore, they can change the rigidity of the cement structure, for instance by blocking the possibility of a rearrangement of the coagulated system. Experiments are in progress to study these influences.

5 Conclusions

The formation of the interfacial transition zone has never been investigated. Based upon the microstructure, a relatively high amount of water has to be present near the aggregates in the first few hours of hydration. Various authors have made remarks as to what could be the cause of this water enrichment. In this paper it is shown that these remarks still have some omits.

Based upon work done at Delft University of Technology, it is thought that syneresis may be an explanation for the water enriched areas near the aggregates. Syneresis leads to exudation of fluid of a colloid system. It is shown by us that cement particles appear to contract and water exudate in hydrating cement paste during the first few hours after mixing.

It is known that the interfacial zone in blast furnace slag cement is substantially smaller than in ordinary Portland cements. In accordance with these observations our experiments reveal that blast furnace slag cement paste expels substantially less water than ordinary Portland cement.

We think that syneresis will also be present with other cementitious systems where the interfacial zone is less extensive than in Portland cement paste, such as is the case for mixtures of Portland cement with puzzolans like silica fume or fly ash.

References

1.   Maso, J.C. (1996) *Interfacial Transition Zone in Concrete; RILEM Report 11*, E & FN Spon, London.
2.   Barnes, B.D., Diamond, S. and Dolch, W.L. (1978) The contact zone between Portland cement paste and glass 'aggregate' surfaces. *Cement and Concrete Research*, Vol. 8, pp. 233-244.
3.   Zimbelmann, R. (1978) The problem of increasing the strength of concrete. *Betonwerk und Fertigteil-Technik*, Vol. 2, pp. 89-96.
4.   Zimbelmann, R. (1985) A contribution to the problem of cement-aggregate bond. *Cement and Concrete Research*, Vol. 15, pp. 801-808.
5.   Strubble, L., Skalny, J. and Mindess, S. (1980) A review of the cement-aggregate bond. *Cement and Concrete Research*, Vol. 10, pp. 277-286.
6.   Diamond, S. (1986) The microstructure of cement paste in concrete. *8th International Congress on the Chemistry of Cement*, Rio de Janeiro, Vol. I, pp. 122-147.
7.   Diamond, S. (1987) Cement paste microstructure in concrete. *Materials Research Society Symposium Proceedings*, Vol. 85, pp. 21-31.
8.   Zimbelmann, R. (1987) A method for strengthening the bond between cement stone and aggregates. *Cement and Concrete Research*, Vol. 17, pp. 651-660.
9.   Scrivener, K.L. and Pratt, P.L. (1986) A preliminary study of the microstructure of cement/sand bond in mortars. *8th International Congress on the Chemistry of Cement*, Rio de Janeiro, Vol. III, pp. 466-471.
10. Hoshino, M. (1989) Relationship between bleeding, coarse aggregate and specimen height of concrete. *ACI Materials Journal*, Vol. 86, pp. 125-190.
11. Mehta, P.K. and Monteiro, P.J.M. (1988) Effect of aggregate, cement and mineral admixtures on the microstructure of the transition zone. *Materials Research Society Symposium Proceedings*, Vol. 114, pp. 65-75.
12. Larbi, J.A. (1991) *Ph. D. Thesis: The cement paste-aggregate interfacial zone*, Delft University of Technology, Delft.
13. Pietersen, H.S. (1993) *Ph. D. Thesis: Reactivity of fly ash and slag in cement*, Delft University of Technology, Delft.
14. Su, Z. (1995) *Ph. D. Thesis: Microstructure of polymer cement concrete*, Delft University of Technology, Delft.
15. Yang, M., Neubauer, C.M. and Jennings, H.M. (1997) Interparticle potential and sedimentation behavior of cement suspensions. *Advanced Cement-Based Materials*, Vol. 5, pp. 1-7.
16. Hunter, R.J. (1993) *Introduction to Modern Colloid Science*, Oxford University Press, New York.
17. Kruyt, H.R. (1949) *Colloid Science, II*, Elsevier Publishing Company, Inc., New York, p. 573.
18. Walstra, P. (1993) *Cheese: chemistry, physics and microbiology, I*, (ed. Fox, P.F.), Chapmann & Hall, London, p. 141.
19. Jefferis, S.A., Sheikh Bahai, A. (1995) Investigaton of syneresis in silicate-aluminate grouts. *Geotechnique*, Vol. 45, pp.131-140.
20. Van Dijk, H.J.M. (1982) *Ph.D. Thesis: Syneresis of curd*, Agricultural University: Wageningen.
21. Nägele, E. (1986) The zeta-potential of cement. *Cem. Conc. Res.* Vol. 16, pp. 853-863.
22. Geiker, M. (1983) *Ph.D. Thesis: Studies of Portland cement hydration by measurement of chemical shrinkage and a systematic evaluation of hydration curves by means of the dispersion model*, Technical University of Denmark, Lyngby.

5 TWO-DIMENSIONAL CONCRETE MODELS USING METAL AGGREGATES

M. SUARJANA and M.S. BESARI
Civil Engineering Department, Institut Teknologi Bandung, Indonesia
R. ABIPRAMONO
Civil Engineering Department, Catholic University of Parahyangan,
Bandung, Indonesia

Abstract
This study was carried out striving at gaining a lucid understanding of the effects of interfacial zones and surface conditions (roughness) of aggregates on the strength of specimens. For this purpose, two dimensional model specimens measuring 100 x 200 x 40 mm were prepared using cement-sand mortar, containing cylindrical steel aggregates. The steel cylinders, which is stronger than the mortar matrix and do not absorb water, were given smooth as well as grooved surfaces, to simulate varying surface conditions. Test results show that addition of small numbers of aggregates into the specimen has a detrimental effect upon specimen strengths. The strength decreases from 70 MPa at specimens with no inclusions to 34 MPa at specimens with 6 smooth surfaced inclusions and regains strength at higher inclusion numbers. Results also indicate that specimens containing rough surfaced inclusions consistently provide higher strength than specimens with smooth surfaced inclusions. At the lowest point, the use of rough surfaced aggregates in the specimen increases its strength by approximately 35 %. Test results are compared against FEM analyses.
Keywords: Concrete model, crushing strength, cylindrical steel inclusion, grooved surfaces, interfacial zones, metal aggregates, model specimen, non-linear stress analysis.

1 Introduction

The general quality of concrete is commonly given as the crushing strength of small specimens made of the same materials and tested at a previously determined age.

The Interfacial Transition Zone in Cementitious Composites, edited by A. Katz, A. Bentur,
M. Alexander and G. Arliguie. Published in 1998 by E & FN Spon, 11 New Fetter Lane,
London EC4P 4EE, UK, ISBN: 0 419 24310 0

The specimens are usually in the form of cubes or cylinders and tested for their strengths at age 28 days. The resulting strength is ordinarily accepted as representing the bulk strength of a much larger amount of concrete being considered. It is well established among concrete professionals that the strength of concrete is the synergetic result of many parameters, such as the quality of the constituent materials, i.e. cement, sand and gravel, the amount of water or the value of w/c ratio used, the method of casting and compaction, the method and the duration of curing and the specimen age at testing [1] [2].

Condensed Silica Fumes (CSF) and Fly Ash (FA) have lately found frequent application in concrete as partial replacement of cement, in particular in preparing High Performance Concretes. Concretes of strengths reaching over 100 MPa have been achieved in laboratory environments as well as in actual construction sites. The prevailing high strength in such concretes is primarily due to quality improvements in the interfacial zones around the aggregates, where minute silica fume particles act as filler as well as induce pozzolanic reactions with lime produced by the hydration process of cement [3] [4].   Bentur et. al. [5] have in their work conclusively demonstrated the beneficial effects of CSF in improving the condition of interfacial zones in concrete.

The advent of powerful Main Frames earlier and later followed by cost effective Personal Computers has in the past 25 years been instrumental in complementing laboratory investigations with mathematical modelling and computer simulations in a wide range of concrete subjects encompassing massively large structures, such as dams and off-shore oil rigs, to structural elements such as columns and beams, and to small laboratory models such as short columns, cylinders and cubes.

In their study, Thorenfeldt et. al. [6] proposed to represent High Strength Concrete behavior by a mathematical model involving ultimate strengths and strains obtained from standard concrete tests.   Chung et. al. [7] also made use of mathematical models for concrete under biaxial stress in their study of shear critical beams. The equation considered was of the fractional type involving peak strength and strain under biaxial stress.

2 Purpose of Investigations

The main purpose of this investigation is to gain a lucid understanding of how the interfacial zones, existing around coarse aggregates, affect the bulk strength of concretes. The relationship between certain relevant aggregate conditions, such as their surface conditions and their strengths, need to be looked into critically in order to find ways to accurately predict the crushing strength and post peak behavior of concrete specimens. It is also hoped that a clear understanding of such relationship may eventually lead to possibilities of proportioning cost effective high performance concretes.

3 The Model Specimen

For investigation purposes, model specimens were constructed such that the governing mechanisms of interest are retained while elements providing undesired effects are eliminated. To simplify prevailing stress conditions, a model capable of producing plane stress conditions is desired. For that purpose, concrete model specimens measuring 100 x 200 mm with a constant thickness of 40 mm, were prepared using cement-sand mortar, where a varied number of steel cylinders, representing coarse aggregates, were included. Thinner specimens would produce stresses closer to a plane stress condition, however, a minimum thickness of 40 mm was deemed necessary to avoid buckling of the specimen during compressive testing.

Steel cylinders were deliberately chosen to provide aggregates stronger than the mortar matrix and which do not absorb water, hence excluding the possibility of across aggregate splitting and eliminating the effects of water released from saturated aggregates. The steel cylinders, all having 20 mm diameters, were given two types of surface conditions. Ones with smooth surfaces and others with tiny circumferential grooves to simulate surface roughness.

4 The Test Specimen

A total of 40 model specimens were prepared and tested. One group of 20 specimens were given steel aggregates with grooved surfaces while the remaining 20 specimens had smooth surfaced aggregates. The aggregates were arranged according to 6 definite patterns. Fig. 1 presents seven model specimens of which six contained varying

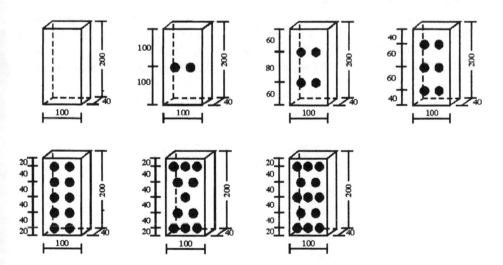

Fig. 1   Model specimens with varying numbers of inclusions, all measures in mm.

Table 1. Mix composition/m³ of cement-sand mortar

Components	Units	
Cement	(kg)	876.371
Sand	(kg)	1 187.570
Water	(kg)	187.332
Superplasticizer	(ltr)	17.527

numbers of inclusions. The model specimens were left in the mould for one full day after casting and   upon demoulding were further cured under water for 26 days.

The cement-sand mortar used in producing the model specimens was prepared according to the mix shown in Table 1. To gain information of mortar mechanical properties, a number of 300 x 150 mm and smaller 200 x 100 mm standard cylindrical specimens were prepared and tested. The specimens were prepared and cured in a way similar to that of the model specimens.

5 Test Results

One day prior to testing, specimens were removed from the curing water and left to dry freely on the laboratory table. The following day, at age 28 days, they were tested for their crushing strengths, on a Universal Testing Machine, applying strain control, such that the softening post peak branch of the stress-strain curve could still be recorded. Table 2 lists the crushing strengths of cylindrical specimens obtained from tests. Fig. 2(a) displays the averaged stress-strain curves of those specimens. These results closely discribe a brittle behavior of high strength mortar, i.e. exhibiting linear behavior for most of the ascending branch of the curve and non-linearity in a small area around the top and immediately descends for a short pass .

The mechanical properties of the steel used to produce cylindrical aggregates were obtained from a tensile test on a 750 mm long cylindrical rod of 20 mm diameter, made out of the same material. Fig. 2(b) shows the resulting stress-strain diagram. It exhibits an elastic limit of f_y = 573 MPa at yield strain ε_y = 0.0134, producing an elastic modulus of E_s = 42761 MPa.

A total of 19 model specimens with smooth surfaced inclusions and 16 model specimens with rough surfaced aggregates were tested. Their averaged results are

Table 2. Crushing strength of mortar cylinders at age 28 days

Number	ID Label	Size (mm)	Crushing stress (MPa)	Ave. stress (MPa)
1	1-4 R	300 x 150	61.546	
2	1-5 R	300 x 150	61.135	
3	1-6 R	300 x 150	63.498	
4	1-9 R	300 x 150	68.757	63.734
5	RA-4	200 x 100	82.262	
6	RA-5	200 x 100	79.945	
7	RA-6	200 x 100	77.527	79.911

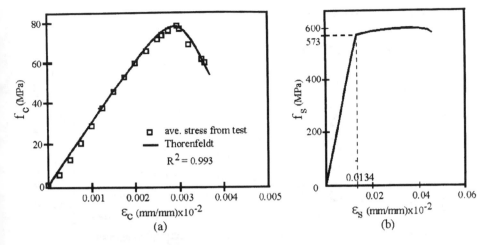

Fig. 2 (a) Averaged Stress-Strain Relation of 100 x 200 mm Mortar Cylinder
(b) Stress - Strain Diagram of Steel

tabulated in Table 3 [8]. Fig. 3 displays curves of their strengths plotted versus their
number of inclusions, which is proportional to the amount of included interfacial zones.
For convenience, the right side of the figure is also used to indicate non-dimensional
stress ratios $\bar{f}_c / \bar{f}_{co}$, where $\bar{f}_c$ represents the crushing strength of model specimens
and $\bar{f}_{co}$ is the averaged crushing strength of specimens containing no inclusions. It
may be observed that for the cases being investigated the crushing strength of
specimens with rough surfaced inclusions are consistently higher than that produced by
smooth surfaced specimens.   Both curves clearly show that at small numbers of
inclusions, i.e. small amounts of interfacial zones, the crushing strengths decrease at

Table 3. Crushing strength of model specimens, age 28 days.

Number	ID Label	Number of specimen	$\bar{fc}$ [1] (MPa)	ID Label	Number of Specimen	$\bar{fc}$ (MPa)
1	ICH 0 [2]	4	69.853 [3]	ICK 0 [4]	3	69.853 [3]
2	ICH 2	3	50.536	ICK 2	2	52.551
3	ICH 4	3	37.430	ICK4	3	50.884
4	ICH 6	2	34.924	ICK 6	2	48.481
5	ICH 10	3	45.309	ICK 10	2	53.738
6	ICH 11	2	50.911	ICK 11	2	63.318
7	ICH 13	2	62.408	ICK 13	2	72.085

1) $\bar{fc}$ = Averaged crushing stress of model specimen.
2) Letter H indicates smooth aggregate surface, trailing digits indicate number of
aggregates included in the specimen.
3) Averaged from seven specimen results.
4) Letter K indicates grooved aggregate surface, trailing digits indicate number of
aggregates included in the specimen.

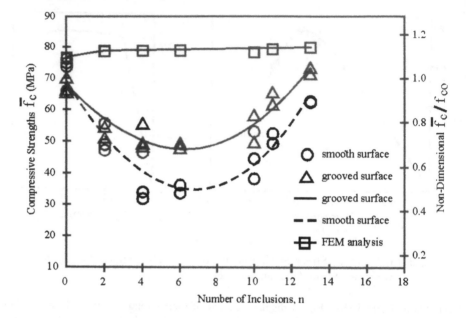

Fig. 3 Model Specimen Strength Versus Number of Inclusions

diminishing rates with increasing numbers of inclusions, reaching their lowest values at approximately 6 inclusions, and start regaining strengths thereafter.   At the lowest points, crushing strengths of 35 MPa and 48 MPa were recorded by specimens containing smooth and rough surfaced aggregates respectively.   Comparing these values with that produced by non aggregated specimens led to strength reductions of 50% and 31% respectively.   At 13 inclusions, the rough surfaced specimens have not only completely regained their lost strength, but have surpassed $\bar{f}_{co}$.   The general trends of both curves strongly suggest the possibility of continued strength growth into values well beyond $\bar{f}_{co}$.

6 Stress Analysis

The decreasing strength at low numbers of aggregates may with some degree of confidence be attributed to the presence of interfacial zones in the concrete, since no other rationally acceptable reason is available. However, strength increase at growing numbers of inclusions, hence increasing weakening interfacial zones, require careful deliberations.   Although the only legitimate reason left would be the presence of stronger steel aggregates, a cautious approach would be commendable.

   To gain a clear understanding of the strengthening phenomenon, the ultimate strengths of the model specimens were simulated through Finite Element Method (FEM) analysis, applying elasto-plastic steel and non-linear mortar properties.   The FEM model assumes a perfect bond between aggregates and the mortar matrix and hence eliminating weakening interfacial zones. To facilitate the process of analysis, the non-linear behavior of mortar was modelled according to a constitutive equation for high strength concrete, as proposed by Thorenfeldt et.al. (1987), however, adapted to fit mortar properties. The equation is of the form :

$$\frac{f_c}{f_c'} = \frac{\varepsilon_c}{\varepsilon_c'} \cdot \frac{n}{n - 1 + \left(\dfrac{\varepsilon_c}{\varepsilon_c'}\right)^{nk}} \tag{1}$$

where for mortar

$$n = 3.8 + \frac{f_c'}{17} \, \text{MPa}, \tag{2}$$

$$k = 0.81 \text{ when } \frac{\varepsilon_c}{\varepsilon_c'} \le 1, \quad k = 0.01 + \frac{f_c'}{65} \text{ when } \frac{\varepsilon_c}{\varepsilon_c'} > 1, \tag{3}$$

f_c and ε_c are the stress and strain of mortar respectively, f_c' is the crushing strength of mortar cylindrical specimen and ε_c' represents the strain of mortar at stress f_c'. Fig. 2(a) displays how well the adapted Thorenfeldt's equation simulates the stress-strain relationship of mortar compared to averaged data obtained from test of 100 x 200 cylinders. The crushing strength achieved by FEM analyses are presented in Table 4, showing consistently higher strengths than those obtained from tests and increasing slowly with growing numbers of inclusions. Hence, pending on results of further research, it may temporarily be speculated that the rise in strength beyond the lowest points is due to the hardening effects of aggregate's superior material and their topologies, exceeding those of the diminishing debilitating tendencies of growing amounts of interfacial zones in concretes.

Table 4. Results of FEM analysis

Number	ID Label	Ultimate Strength (MPa)
1	ICF0 [1]	76.579
2	ICF2	78.691
3	ICF4	78.791
4	ICF6	78.883
5	ICF10	78.259
6	ICF11	79.342
7	ICF13	79.785

[1] Letter F stands for FEM model. trailing digits represents number of inclusions

7 Concluding Remarks

Results of this investigations, covering the cases considered in this study, lead to the following conclusions :

1. The addition of small numbers, up to six or seven, of steel aggregates decreases the model strength at a diminishing rate down to the minimum point.
2. Beyond the minimum point, the strengthening effects of aggregate's superior material and their topologies exceed the diminishing detrimental effects of weak interfacial zones.
3. Rough surfaced aggregates improve the crushing strengths of the model specimens.
4. Thorenfeldt's Equation simulates the Constitutive Equation of mortar accurately, leading to excelent results of ultimate strength by non-linear FEM analysis.
5. Absence of perfect bond between aggregates and mortar lead to lower model specimen strengths obtained from actual tests.
6. Further research considering varying aggregate strengths and topologies still needs to be conducted to conclusively determine their effects on the strength of the concrete model.

Acknowledgement

The authors gratefully acknowledge the partial support from Directorate General of Higher Education through Team Research Project Contract No. 016/HTPP-III/URGE/1997. Appreciation is also due to DR. D.R. Munaf, Chairman of the Structures and Material Laboratory of Institut Teknologi Bandung, where most of the investigation work was carried out, for his generous assistance.

References

1. Troxell, G.E., Davis, H.E. and Kelly, J.W. (1968), *Composition and Properties of Concrete*, McGraw-Hill, New York.
2. Neville, A.M. (1990), *Properties of Concrete*, ELBS, Singapore.
3. Besari, M.S., Munaf, D.R., Hanafiah (1992), The Effect of Fly Ash and Strength of Aggregate to Mechanical Properties of High Strength Concrete, *Proc. Conf. Conc. and Struct.*, Singapore 25-27 August 1992.
4. Larbi, J.A., (1993), Microstructure of Interfacial Zone around Aggregate Particles in Concrete, *Heron*, Vol. 38, 1993, No. 1, Netherland.
5. Bentur, A., Goldman, A., and Cohen, M.D., The Contribution of The Transition Zone on The Strength of High Quality Silica Fume Concrete, Mat. Res. Soc. Symp. Proc. Vol. 114, 1988, *Material Research Society*, pp. 97 - 105.
6. Thorenfeldt, E., Thomaszewics, A. and Jansen, J.J. (1987), Mechanical Properties of High Strength Cncrete and Application in Design, *Proc. Symp. Utilization of High Strength Concrete*, June 1987, Norway.
7. Chung, W., Ahmad, S.H., (1994), Model Shear Critical High Strength Concrete Beams, *Struc. Jour., ACI*, January 1994.
8. Abipramono, R., (1997), The Effect of Amount and Distribution of Aggregates on The Strength of Two Dimensional Concrete Models, Master Thesis, Civ. Eng. Dept., Institut Teknologi Bandung (in Indonesian).

6 INTERFACIAL TRANSITION AND DESTRUCTION IN HARDENING OF WATER–SILICATE DISPERSION SYSTEMS

D.I. SHTAKELBERG and S.V. BOIKO
The Standards Institution of Israel, Tel-Aviv, Israel

Abstract

The major causes of destruction during the hardening process are interfacial transitions of a colloidal-chemical nature, taking place at the weakest spots of the newly formed structures: in the zones of coagulational contacts between dispersed particles (particle aggregates), products of the chemical reactions of hydration and hydrolysis.

A thermodynamic analysis of the interaction between capillary and surface forces, localized in the adsorption films of liquid moisture between the solid particles, has been made, based on the balance of change of the free energy of Helmholz.

It has been demonstrated that the spontaneous release of free energy resulting from the disruption of exceedingly thin (1 to 3 molecular levels) adsorption films are a major cause of the formation of primary microcracks and the temporary loss of strength in hardening cement composites.

Keywords: Coagulational contact, destruction, disruption, hardening, interaction, microcrack, pellicular moisture

1. Introduction

The process of hardening of water-silicate dispersion systems (cement compositions, in the first place) should be regarded as a series of sequential transitions of the material from one structural state to another: the initial dispersion system (more or less concentrated suspension) → distant coagulation structure (colloid capillary-porous

The Interfacial Transition Zone in Cementitious Composites, edited by A. Katz, A. Bentur, M. Alexander and G. Arliguie. Published in 1998 by E & FN Spon, 11 New Fetter Lane, London EC4P 4EE, UK, ISBN: 0 419 24310 0

body) → close coagulation structure (solid-like capillary-porous body) → condensation or condensation-crystalline structure (solid capillary-porous body).

At the same time, the hardening of binding systems is, by its very nature, an alternative process of the "structure-formation-destruction" type [1,2] where the development of structural stresses is due to the action of either internal chemical and physicochemical factors or the external, i.e. technological-mainly temperature-moisture influences. The appearance of the volume-stressed state in dispersion systems results in the appearance of a network of initial microcracks whose enlargement brings about the reduction of strength of the material and even the complete loss of its load carrying capacity.

The structure-formation of watered dispersions, just like any other phenomenon of the material world, must be studied from the point of view of cause and effect. In the case under consideration, it means that the weak zones of the structure under formation which are the most susceptible to crack formation, must be detected and the reasons, i.e. the forces causing microdestruction in these places need to be found.

Lt us put aside the phenomena of a purely chemical nature (increase in the volume of reaction products, exothermic effects, etc.) as well as external technological influences (heat curing and the thermal stresses caused by it, excess internal pressures due to the gaseous phase expansion, diffusion processes, etc.). Let us turn to the physicochemical aspects of the problem directly connected with the dehydration of dispersion systems and, in particular, to the properties of the pellicular moisture which determines the physical and mechanical state of the dispersion systems concentrated up to the limit.

2. Contacts in hardening structure

In this connection, what requires consideration is the nature of the interaction between solid-phase particles and their aggregates which are determined by the state and the properties of the dispersion medium in the system.

The strongest crystallization contacts (Fig. 1, a) formed by forces of a chemical nature (coordination and covalent bond with an energy in the order of 20 kcal/mole) reach strength factors (up to 1500 MPa) compatible with the theoretical strength values of hardened cement paste and concrete [3,4].

The strength of the cement specimens formed and hardened under super high pressures (up to 700 MPa) and high temperatures approaches these values [5]. The conventional kinetic laws of hardening are already invalid under such conditions, and coagulation contacts practically fail to form since the water molecules become mechanically and directly embedded in the crystal lattices of cement materials.

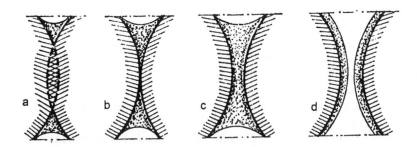

Fig. 1. Scheme of various types of contacts in dispersion systems: a) solid-phase crystallization contact; b) solid-phase coagulation contact; c) coagulation contact through a water layer; d) coagulation contact through an air space.

Extremely low (below 0.05) porosity values of hot pressed specimens also testify to this fact.

However, the strength of the hardening (and hardened) water-silicate dispersion systems produced in accordance with conventional technological patterns is 1.5 to 2 orders of magnitude lower than the theoretically calculated value. Consequently, it can be assumed that contacts of chemical nature do not influence substantially the macroscopic strength of such materials. It means that the destrucction of the hardening and drying structure of silicate dispersions does not occur in the solid phase but takes place in the weaker contacts of close (Fig. 1, b) and distant (Fig. 1, c, d) coagulations.

The close-action coagulation contacts, or "dry" mineral contacts according to S.V. Nerpin [6] (Fig. 1, b), are formed due to hydrogen bonds whose energy is 5-8 kcal/mole. Since the length of hydrogen bonds is only 1-3Å, which is smaller than the water molecule diameter, these contacts represent direct solid-state (but not chemical!) interactions with no participation of the liquid phase. Theoretically, the strength level in the order of 50 MPa can be achieved by hydrogen bonds, and, evidently, it may be further increased by the compacting action of capillary forces.

The strength of distant coagulation contacts (Fig. 1, c, d) is substantially lower since the contacts are formed by distant-action electrostatic and dispersion forces of van der Waals-London with a bond energy of less than 1 kcal/mole.

3. Interfacial transition and destruction

The role of the liquid phase in such contacts is extremely great and, at the same time, peculiar: here, the physicochemical moisture is a rightful component of the capillary-porous structure which determines directly the variations in the strength of the material.

Thus, if the distance between the opposite solid surfaces is sufficiently short and if their adsorptive layers overlap (Fig. 1, b, c), then, according to B.V. Deriagin [7], the action of the adsorptive component of the wedging pressure begins, and its positive value may exceed the molecular attraction which will bring about mutual repulsion of the two solid surfaces and the loss of strength of the contact. When the liquid is removed by way of chemical binding or diffusion, the solid surfaces once again approach each other, thus increasing the level of solid-state interaction and consequently, the local strength value. Then the cycle is repeated, etc.

This mechanism of change in the structure and moisture of dispersion systems makes it possible to explain the wave-like changes of strength observed in the course of hardening of mineral binding materials.

Further development of drain in the mass of moisture leads to the disruption of its continuity in the coagulation contact zones (Fig. 1, d). The system becomes unstable or metastable, the increase of strength temporarily stops and sometimes even reductions of strength occur. However, this creates preconditions for the evolution of the (1, d) contact type which turns into the (1, b) type bringing about a pronounced strengthening of the system. A similar effect is observed both in the chemically active (hardened cement paste, concrete) and in the chemically inert (clay) water-silicate dispersions [8].

The continuity disruption of the liquid phase and its transition into a pellicular adsorptive state do not involve the cessation of the mechanical influences on the solid surface. On the contrary, these influences increase.

To prove the aforesaid, let us consider the structure element (Fig. 2, a) which is a bent solid surface (phase 3). Let us assume, for the sake of simplicity, that this surface is part of a sphere with a radius r on which there is an adsorptive layer of liquid (phase 2) with thickness, h_2, interacting with the surrounding gaseous medium (phase 1).

The drying adsorptive pellicle of liquid tends to the energetically most beneficial state, i.e. it tends to change over to a state with the least dividing surface area 2, 3. In practice, it means that, due to the capillary, superficial and other effects, the adsorptive layer is submitted to the resulting compressing pressure, P, whose tangential component, P_t, is transmitted to the resilient plastic solid surface by the adhesion forces and deforms (compresses) it.

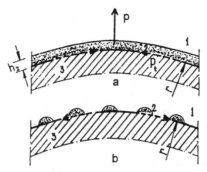

Fig. 2. Changes of the state of a dispersion structure element in the dehydration process: a) state with pellicular moisture; b) state with drop-like moisture.

4. Thermodynamic analysis

The value of pressure, P, can be determined by the thermodynamic method from Helmholtz's free energy modification balance which, in the case under consideration, is represented by the following:

$$dP = -PdV_2 - \sigma_{12}d\omega_{12} + \mu_2dn_2 \qquad (1)$$

or

$$\frac{dF}{dV_2} = -P - \sigma_{12}\frac{d\omega_{12}}{dV_2} + \mu_2\frac{dn_2}{dV_2} \qquad (2)$$

where P is the pressure of mechanical work done by the system; V_2 is the volume of the adsorption layer, σ_{12} is the surface tension coefficient on the "liquid-gas" dividing surface; ω_{12} is the area of this surface; μ_2 is the chemical potential of the liquid; and n_2 is the number of moles in the liquid.

The volume variation of free energy is the pressure taken, with the reverse sign. Consequently, the left part of equation (2) determines the value of general pressure, P, in the adsorptive layer; the second term in the right part of equation (2) is the capillary pressure, P_L, resulting from the curvature of surface 12, and

$$p = P + P_L - \mu_2\frac{dn_2}{dV_2} \qquad (3)$$

The third term in equations (2) and (3) is directly linked to the liquid layer energy when its mass and volume diminish. Since

$$\mu_2 = \left(\frac{\partial F}{\partial n_2}\right)_{T,V,n_i \neq n_2,} \tag{4}$$

then

$$\left(\frac{\partial F}{\partial n_2}\right)_{T,V,n_i \neq n_2} \cdot \frac{dn_2}{dV_2} = -P_2, \tag{5}$$

where P_2 is the "concentration" pressure in the adsorptive layer.

The above relationship, (5), determines the state of the liquid provided the thickness values, h_2, are sufficiently high, the pressure in the layer still conforms to the law of Pascal, and the liquid possesses volumetric properties.

If the thickness of the adsorptive layers is diminished to reach the value of 100 Å and less, the liquid acquires anomalous properties (increased density and viscosity, reduced dielectric permeability) due to the influence of the solid surfaces.

Then, expressing the layer volume as $dV_2 - \omega_{23}dh_2$, we obtain from expression (5):

$$\left(\frac{\partial F}{\partial n_2}\right)_{T,V,n_i \neq n_2} \cdot \frac{dn_2}{\omega_{23}dh_2} = \frac{df_{23}}{dh_2} = -\Pi, \tag{6}$$

where df_{23} is the free surface energy, Π, is the wedging pressure defined according to [7] as:

$$\Pi = P_{23} - P_2, \tag{7}$$

where P_{23} is the pressure on the "liquid-solid phase" dividing surface.

Consequently, the balance of pressures (3) may be written for the volumetric layer as:

$$p = P + P_L + P_2 \tag{8}$$

and for the thin pellicle as

$$p = P + P_L + \Pi, \tag{9}$$

while, in both cases, the general pressure, P, increases since $dV_2 \rightarrow \min$ (2).

The reduction of the pellicle thickness occurs only up to a certain minimal critical value, $h_{2(cr)}$, which, on reaching it, the pellicle breaks, disintegrating into individual "lenses" [6] (Fig. 2, b). As this takes place, the pressure, P, falls momentarily from its maximum value to zero, and the previously compressed solid surface is "unloaded" in the same momentary manner by extending under the influence of elastic forces. However, the elastic properties of the material which has not yet undergone final hardening, erode with an extremely high speed, and the deformations that follow develop already in the plastic field which is unable to absorb the dynamic influences resulting from the formation of local ruptures on the surface in places weakened by foreign inclusions and other defects.

5. Conclusion

Thus, it becomes possible, within the framework of the general provisions of the interfacial interaction theory, to refine our understanding of the formation of macroscopic strength of water-silicate dispersion systems, on the one hand, and to reveal new regularities in the action of destructive phenomena in the course of the hardening process.

6. References

1. Segalova, E.E. and Rebinder, P.A. (1960) Modern physicochemical insights into the hardening processes of mineral binding substances. *Stroitelnyie materialy* (Building materials), No. 1, pp. 21-26.
2. Shtakelberg, D.I. and Rukmane, Y.Y. (1982) Ultrasonic control of ceramics in the drying process. *Steklo i keramika* (Glass and ceramics), No. 7, pp. 21-23.
3. Akhverdov, I.N. (1981) *Principles of the Physics of Concrete*, Stroiizdat, Moscow.
4. Krylov, N.A., Kalashnikov, V.A. and Polishchuk, A.M. (1966) *Radiotechnical Methods of Quality Control of Reinforced Concrete*, Stroiizdat, Leningrad-Moscow.
5. Roy, D.M. and Gouda, G.P. (1976) Strength optimization of the cement test, *Sixth International Congress on Cement Chemistry*, Vol. 2, book 1, Stroiizdat, Moscow, pp. 310-315.
6. Nerpin, S.V. and Chudnovsky, A.F. (1967) *Soil Physics*, Nauka Publishers, Leningrad, pp. 548-552.
7. Deriagin, B.P. (1979) Superficial forces and the wedging pressure, *Physical Chemistry of the Surface*, A. Adamson, Mir Publishers, Moscow.
8. Shtakelberg, D.I. (1984) *Thermodynamics of Structure Formation of Water-Silicate Dispersion Materials*, Zinatne Publishers, Riga.
9. Shtakelberg, D.I. and Sychev, M.M. (1990) *Self-Organization in Dispersion Systems*, Zinatne Publishers, Riga.

PART THREE
ROLE OF ITZ IN CONTROLLING DIFFUSION AND PERMEABILITY

7 GAS PERMEABILITY OF MORTARS IN RELATION WITH THE MICROSTRUCTURE OF INTERFACIAL TRANSITION ZONE (ITZ)

M. CARCASSES, J.Y. PETIT and J.P. OLLIVER
Laboratoire Materiaux et Durabilité des Constructions, INSA-UPS,
Toulouse, France

Abstract

Permeability of concrete may depend on the aggregate content because :
- the permeability of the two components, cement paste and aggregate, are different,
- the tortuosity of the porous paths, the volume fraction of ITZ and their connectivity increase with the aggregate content.

Gas permeability of cement based materials depends on the microstructure and also on the degree of saturation. The influence of ITZ on gas permeability has been studied on mortars specimens with different sand volume fractions. By conditioning the specimens prior to the gas permeability measurements, the water distribution into the material is modified. Following the Kelvin Laplace equation and assuming an equilibrium, the conditioning treatment implies a maximum pore size d_m to be full of water. If the tested cement based material contains an interconnected pore system which dimension is greater than d_m, the gas permeability should be affected. In that sense, gas permeability is a test which allows us to detect the evidence of an interconnected pore system greater than d_m. If the pore size is greater than d_m in ITZ, gas permeability data should indicate the connectivity of ITZ pores. An experimental program has been conducted with two W/C ratios (W/C = 0.5 and 0.35). With the lowest W/C ratio, two types of specimens have been tested : a first series with an OPC as binder and a second series with OPC and silica fume.

The interconnection of ITZ can be analysed by means of a geometrical model. With the sand gradation and assuming an ITZ thickness of 20 µm, the sand volume fraction corresponding to a percolation threshold (interconnection of ITZ) is around 45%. Experimental data on cement based materials cast with W/C = 0,50, show that below this threshold, gas permeability decreases with the sand content whereas it increases above 45% : this result can be analysed in terms of ITZ effect. With a low water to cement ratio, gas permeability decreases regularly with the sand content : the influence of ITZ appears as very little in this case.

Keywords : gas permeability, ITZ, model, sand volume fraction.

The Interfacial Transition Zone in Cementitious Composites, edited by A. Katz, A. Bentur,
M. Alexander and G. Arliguie. Published in 1998 by E & FN Spon, 11 New Fetter Lane,
London EC4P 4EE, UK, ISBN: 0 419 24310 0

1 Introduction

It is generally assumed that transport properties of cement based materials are affected by the microstructure of the Interfacial Transition Zone (ITZ). This assumption stands on the characteristics of the porous system in the cement which is influenced by aggregates as shown figure 1. For mature cement based materials cast with a water to cement ratio equal to 0.4, the ITZ is about 3 times more porous than the bulk cement paste[1]. When the aggregate volume fraction increases, the degree of interconnection of ITZ increases and a threshold effect appears depending on the grain size distribution and the ITZ thickness.

Diffusion coefficient of chloride ions has been measured in mortars for different sand volume fractions.[2] The mortar/neat paste diffusion coefficient ratio has been found to decrease almost linearly with the increase of the sand volume fraction even though all the ITZ appear to be interconnected. The results of the same study did not indicate the existence of a sand volume threshold from which the overall diffusion coefficient increases rapidly. This result on chloride diffusion can be compared with the model developed by Garboczi et al.[3] for DC conductivity. For these two transport phenomena, the flow is proportional to the porosity of the solid and independent of the pore size. Consequently, the inconnection of the larger pores belonging to the ITZ percolation cluster for a sufficient sand volume fraction has few consequence on the flow.

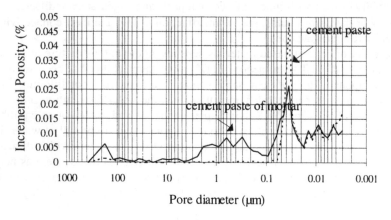

Fig. 1. Incremental porosity distribution of pure cement paste (W/C = 0.4) and cement paste of mortar.

On the contrary, such an effect can be expected for other transport properties depending on the pore size, i.e. the permeability.

2 Materials and sample preparation

We have studied two W/C ratio mortar series. A first one with a W/C ratio of 0.5, and another one with 0.35. For each series, the sand volume fraction varies from 10% to 60%. The binder is an OPC (CEMI 52.5R). The sand is a 0.6/2.5 mm roller river aggregate which grain size distribution is shown on figure 2. A third series has been cast with a W/C ratio of 0.35 and silica fume (10% of the binder content : silica fume + cement). Superplasticizer content has been adjusted in order to have the same consistency than the reference mortar cast with W/C = 0.50. The mix compositions are given in tables 1 to 3.

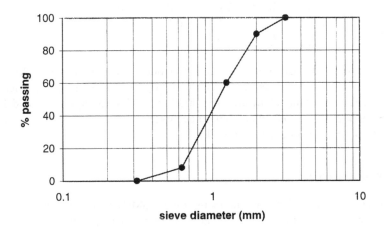

Fig. 2. Sand grain size distribution

Table 1. Mix proportions of W/C=0.5 mortars

sand volume fraction	10 %	19 %	30 %	37 %	40 %	45 %	50 %	57 %
cement (kg/m^3)	1097	987	853.3	767.8	731.3	670.4	607.8	524
sand (kg/m^3)	267	507.6	801	987.8	1068	1201.4	1311	1522
water (kg/m^3)	548.5	493.5	426.5	383.8	365.8	335.2	304	262

Table 2. Mix proportions of W/C=0.35 mortars

	19 %	30 %	45 %	50 %	57 %
sand volume fraction					
cement (kg/m^3)	1208	1044	820.3	745.7	641.3
sand (kg/m^3)	507.3	801	1202	1335	1521.8
water (kg/m^3)	422.8	365.4	287.1	261	224.5
superplasticizer (%C)	0.5	0.86	1	1.2	1.5
entrapped air (%)				2.4	3.4

Table 3. Mix proportions of W/C=0.35 + SF mortars

sand volume fraction	19 %	30 %	45 %	50 %	52 %	57 %
cement (kg/m³)	1087.2	939.5	738.2	671.2	644.3	577.2
silica fume (kg/m³)	120.5	104.3	82	74.5	71.5	64.2
sand (kg/m³)	507.3	801	1202	1335	1388.3	1521.8
water (kg/m³)	422.8	365.4	287.1	261	250.5	224.5
superplasticizer (%C)	0.5	0.5	0.5	1.2	1.3	1.5
entrapped air (%)	2.6	2.4	4.4	5	4.1	4.2

Fresh concrete is cast into (ϕ=15 cm, h=5 cm) cylinders. After one day in a 20°C 100% RH room, the specimens are demolded and then kept for 2 days in a controlled room at 20°C 50% RH. After then, they are oven dried at 50°C for 3 days and stored 1 day in a dessiccator at 20°C for the temperature to decrease.

For each series, 3 specimens have been tested.

3 Permeability measurement

The apparatus used for permeability measurement is the CEMBUREAU permeameter. This is a constant head permeameter using oxygen as the permeating fluid. The apparatus is shown on figure 3.

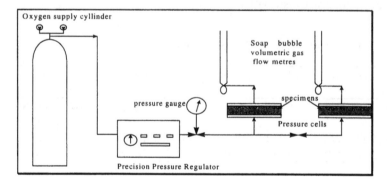

Fig. 3. CEMBUREAU permeameter

A pressure of 3 bars (0.3 MPa) is applied to the inlet face of the specimens. Samples are cylinders of 15 cm diameter and 5 cm thick ; their lateral surface is sealed by a rubber ring pressing under high pressure against the curved surface. The volume flow rate through the specimen is measured with a soap bubble flow meter.

When the steady state flow is reached, by using Hagen-Poiseuille equation for laminar flow of a compressible fluid, the permeability K of the sample is given by :

$$K = \frac{2QP_{atm}L\mu}{A(P^2 - P_{atm})}$$

where :

K = coefficient of gas permeability (m²)
Q = gas flow (m³/s)
A = cross-sectional area (m²)
L = specimen thickness (m)
μ = coefficient of viscosity of the gas (N.s/m²)
P_{atm} = atmospheric pressure (N/m²).
P = inlet gas pressure (N/m²).

4 Gas permeability : experimental data

Gas permeability of cement based materials depends on the microstructure and on the degree of saturation. By conditioning the specimens prior to the gas permeability measurements, the water distribution into the material is modified. In the experiments, the mortar specimens have been stored in an oven at 50°C in which the relative humidity is 12%. Following the Kelvin Laplace equation and assuming an equilibrium, this treatment implies some pores to be empty. A rough calculation gives $d_m = 4 \ 10^{-9}$ m as the maximum pore size full of water. In that conditions, if the tested cement based material contains an interconnected pore system which dimension is greater than d_m, the gas permeability should be affected. In that sense, gas permeability is a test which allows us to detect the evidence of an interconnected pore system greater than d_m. In ITZ, the pore size is greater than d_m and, consequently, gas permeability should indicates the connectivity of ITZ pores assuming that the moisture equilibrium is reached. The experimental results are reported in table 4 and figures 4 and 5. K_{mean} is expressed in 10^{-16} m² and is the mean value of 3 specimens, the columns « variation » indicate the coefficient of variation of the 3 measures in per cent.

Table 4. Permeability results

Sand volume fraction %	W/C = 0.5		W/C = 0.35		W/C = 0.35 + SF	
	K_{mean} (10^{-16} m²)	variation (%)	K_{mean} (10^{-16} m²)	variation (%)	K_{mean} (10^{-16} m²)	variation (%)
10	4.38	2	/	/	/	/
19	2.21	6	1.73	38	2.22	26
30	1.31	9	0.41	45	0.69	34
37	1.06	10	/	/	/	/
40	1.03	2	/	/	/	/
45	0.54	15	0.16	1	0.157	63
50	0.79	7	0.096	34	0.24	76
52	/	/	/	/	0.09	27
57	0.88	14	0.095	22	0.068	6

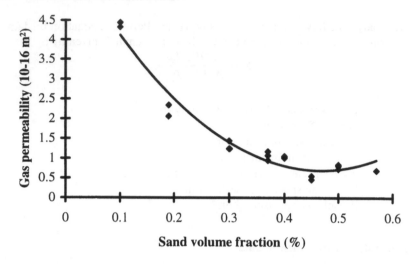

Fig. 4. Influence of sand volume fraction on gas permeability of mortars (W/C = 0.5)

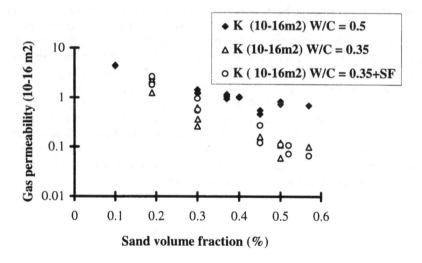

Fig. 5. Influence of sand volume fraction on gas permeability of mortars, effect of W/C ratio and silica fume

5 Analysis of the results and conclusions

The interconnection of ITZ can be analysed by means of a geometrical model[1]. The sand volume fraction corresponding to a percolation threshold (interconnection of ITZ) depends on the sand gradation and the thickness of the transition zone. Taking into account the sand grain size distribution, and assuming spherical grains, percentage of connected ITZ is represented in figure 6 as a function of the sand volume fraction for several realistic ITZ thickness (5μm, 20μm and 30μm). Whatever their thickness, ITZ are connected for sand volume fractions greater than 55%.

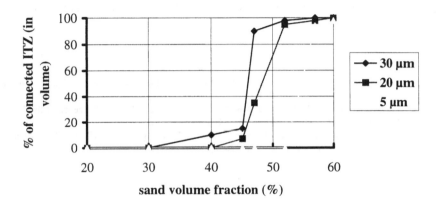

Fig. 6. % of connected ITZ as a function of sand volume fraction

The calculation made by Garboczi et al.[3] shows that the electrical conductivity of mortars decreases regularly with the sand volume fraction. As mentioned earlier, this model has been experimentally confirmed by chloride diffusivity measurements. For gas permeability, the experimental results we present are more contrasted. In mortars cast with a low water to cement ratio (with silica fume or not), the gas permeability decreases regularly with the sand volume fraction (see figure 5).

Assuming a constant pore structure of the paste in mortars cast with a given water to cement ratio whatever the sand fraction, the permeability should decrease if the sand content increases. This result can be expected according to two main facts :
- the sand grains used in this study are impermeable, then the volume fraction of the permeable components decreases with the sand content,
- the tortuosity of the channels in which gas can flow increases with the sand content.

This analysis can be applied for cement based materials prepared with the lower water to cement ratio (W/C = 0.35).

On the contrary, with a higher W/C ratio of 0.5, the gas permeability stabilises for sand volume fraction higher than 0.4. In this case, the stabilisation of the gas permeability can be analysed by a modification of the pore geometry accessible to oxygen during the measurement. Coarser pores belonging to ITZ become interconnected for higher sand fractions and, due to the conditioning conditions of the specimens, it can be assumed that this pore system is not only interconnected but empty of water too. In that conditions, gas permeability ceases to decrease.

In the former case it can be concluded that the pores system in ITZ is not organised in such a way that it allows the gas flow to increases. That means that whether the pore size is not modified in ITZ for low W/C ratio cementitious materials or whether the pores are too small to be dried during the conditioning process.

From this experimental study, it can be concluded that the influence of ITZ on gas permeability of high performance cement based materials is very small. With higher W/C ratio around 0.5, the conclusion is not so clear : permeability stabilises when the aggregate content increases over the percolation threshold of the transition zones.

[1] B. Bourdette, E. Ringot and J.P. Ollivier (1995) Modelling of the transition zone porosity. *Cement and Concrete Research,* Vol. 25, N°4, pp. 741-751.

[2] A. Delagrave, J.P. Bigas, J.P. Ollivier, J. Marchand and M. Pigeon (1997) Influence of the Interfacial Zone on the Chloride Diffusivity of Mortars. *ACBM Journal,* 5, pp.86-92.

[3] E.J. Garboczi, L.M. Schwartz and D.P. Bentz (1994) Modelling the D.C. Electrical Conductivity of Mortars. MRS Symposium proceedings, Vol. 370, pp. 429-436.

8 MODELING THE MODIFICATION OF CHLORIDE PENETRATION IN RELATION TO ITZ DAMAGE

R. FRANÇOIS, G. ARLIGUIE and A. KONIN
LMDC UPS Génie Civi, 31077 Toulouse cedex 4, France

Abstract
This paper deals with the effect of Interfacial Transition Zone damage due to loading on chloride penetration in reinforced high performance and normal concrete. The quantification of ITZ damage is related to the tensile stress in re-bar and then the modification of diffusion is modeled in relation to this tensile stress.
Keywords: ITZ, chloride, diffusion coefficient, damage, tensile strength

1 Introduction

This study is an attempt to characterize the influence of concrete microcracks due to the service load of reinforced structure on the chloride diffusivity. As a result, this concrete damage which is basically located at the ITZ of concrete could lead to a significant effect on long-term prediction of chloride profiles in reinforced concrete structures. To understand and model this effect, an experimental program was built using reinforcing ties made with Ordinary Concrete, High Strength Concrete and Very High Strength Concrete. These ties were stored in a chloride environment over one week periods of wet-humid cycles in a salt fog enclosure. The main originality of this experimental program is that the ties are stored in tensile state, i.e. both cracks and micro-cracks are still opened during the chloride ingress. In this paper, we first characterize concrete damage due to loading, then secondly we analyze the chloride diffusivity in the different concretes to finally propose a model for the increase in diffusivity in relation to load level.

The Interfacial Transition Zone in Cementitious Composites, edited by A. Katz, A. Bentur,
M. Alexander and G. Arliguie. Published in 1998 by E & FN Spon, 11 New Fetter Lane,
London EC4P 4EE, UK, ISBN: 0 419 24310 0

2 Experimental program

Because previous studies [1][2] have shown that ITZ was damaged in the tensile zone of reinforced concrete elements, the experimental program was conducted on reinforced ties stored in loaded tensile state and made of Ordinary Concrete, High Strength Concrete and Very High Strength Concrete. A full description of the experimental program is available in previous paper [3].

3 ITZ damage

3.1 Test method

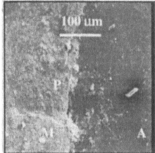

The progression of microcracking was studied by the replica technique [4] and quantified by total projections. Two essential parameters of the microcracking were recorded: firstly, the specific area quantifying the significance of the microcracking network and secondly, the degree of orientation quantifying the microcracking anisotropy. Two other essential parameters for the microcracking could not be recorded by this method: on one hand, the microcrack width and on the other hand the degree of connectivity of the microcrack network.

Fig. 1. A microcrack in the paste

Digitizing the microcrack network is not currently an automatic process. Microcrack extraction is implemented by the researcher working with the SEM. Thus, this is a subjective process which can lead to disparities between different researchers'results. The problem is not crucial for the paste microcracks because these cracks are very obvious (Fig. 1) but it is crucial for ITZ microcracks because these cracks are difficult to separate from the border between paste and aggregates (Fig. 2). As a result, the polishing of samples (necessary before using the replica technique) involves a different step between aggregates and paste because of their different abrasiveness characteristics. This step is inverted by the replica technique and appears as a white border on SEM due to a pin effect increasing electron emission. Furthermore, the degree of polishing can modify the appearance of the edge between paste and aggregate or between paste and re-bar.

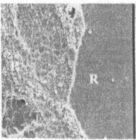

Fig. 2. Paste (P) aggregate (A) or paste (P) re-bar (R) interface for unloaded OC sample (polished until 5 μm as grain size). Is it damaged or not?

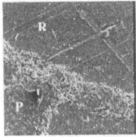

Fig. 3. Paste (P) aggregate (A) or paste re-bar (R) interface for cracked OC sample (polished until 14 μm as grain size). Is it damaged?

For example, Fig. 3 and Fig. 4 show the significance of the degree of polishing on the paste-aggregate interface appearance.

To avoid this problem of subjective extracting of microcracks, we decided to observe on SEM the same area of the sample at different loading levels [5]. Thus, using the comparison between the views located at the same place, we can evaluate the real damage at the paste-aggregate interface.

For example, Fig. 4, Fig. 5 and Fig. 6 show the evolution at the edge between paste and aggregate for a VHSC when not loaded, loaded at pre-cracking state and loading at post-cracking state, respectively. The edge becomes whiter from Fig. 4 to Fig. 6, this phenomenon is significant of the increase in ITZ damage. Obviously, mere observation of Fig. 6 does not entitle us to assert that the interface was damaged.

Because of the heterogeneity of the microcrack network, the method presented above could not characterize the entire microcracking state. The density of microcracking thus calculated provides only local information indicating microcracking susceptibility.

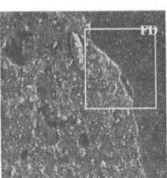
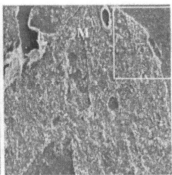

Fig. 4. No damage at the Interface (ND)

Fig. 5. Initial damage at the Interface (FD)

Fig. 6. Increase in interface damage and microcrack in the paste (M)

3.2 Experimental results

For each specimen tested, an area liable to crack during loading was selected. This area was observed in its initial state to provide a reference and was then observed at two loading levels: before and after the macrocrack formation.

For ordinary concrete (OC): results show that there is no initial microcracking in the paste or at the paste-aggregate interface (the same results were reported in a previous work). Increasing load leads firstly to an increase in the density of microcracking basically located at the ITZ to reach 0.45 mm^{-1} and secondly to the formation of a macrocrack and to an increase in the density of microcracking to reach 1.16 mm^{-1}.

For high strength concrete (HSC): results show that there is an initial microcracking due to self dessication [6] (0.50 mm^{-1}). These microcracks are located both in the paste and at the interface between a large grain and the paste. Increasing load leads firstly to a small increase in the length of ITZ damage around the large aggregate to reach 0.54 mm^{-1} and secondly to both the formation of a macrocrack and to an increase in the density of microcracking to reach 1.09 mm^{-1}.

For very high strength concrete (VHSC): results show that there is no initial microcracking in the paste or at the paste-aggregate interface. Obviously, this result does not mean that there is no self-dessication microcracking.

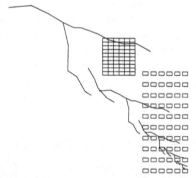

As previously mentioned, because of the heterogeneity of microcracking network, the local zone may not be affected by microcracking
Nevertheless local observation is used in this work to compare the change for a given area in relation to the loading level.

Fig. 7.   Schematic comparison between statistical observation of microcracking network and local observation as used in this work

Increasing load leads firstly to an increase in the density of microcracking basically located at the ITZ to reach 0.76 mm^{-1} and secondly to the formation of a macrocrack and to an increase in the density of microcracking to reach 2.0 mm^{-1}.
Results are summarized on Fig. 8.

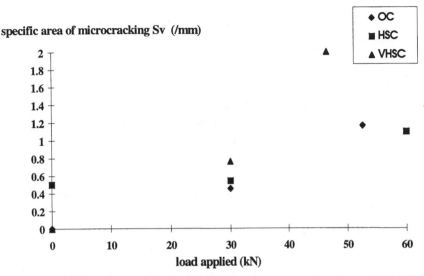

Fig. 8. Variation of the specific area of microcracking for each concrete with loading

As a result, we can conclude that there is an increase of microcracking in relation to the loading level. Whatever the concrete, the damage is mainly located at the ITZ between paste and aggregate. This result was not a priori obvious, because it is generally assumed that the use of silica fume to produce HSC and VHSC leads to a resorption of the interfacial transition zone. Because of the significant variation of Sv in relation to the loading level and because Sv is a local information in this study, the change in Sv in relation to the load applied cannot be modeled.

4 Chloride penetration

Although diffusion is not the only mechanism of chloride ingress into concrete, the usual way to characterize chloride penetration is to calculate the chloride diffusivity value D_a by applying a solution of the Fick's second law according to experimental concentration profiles.

Diffusion is due to a gradient of concentration which can be described in the x-direction by the Fick's first law:

$$J = -D_F \frac{\partial C}{\partial x}$$

where $C(x,t)$ is the chloride concentration kg/m3 of the interstitial solution, J is the flow and D_F the diffusion coefficient.

By taking into account the mass conservation law, binding of a part of chlorides on cement hydrates ($C_t = C_f + C_b$), the porosity p of the material, the dry concrete specific weight ρ_s and the fact that experimental concentration is measured in kg/m^3 of concrete, the equation for Fick's first law leads to Fick's second law [7]:

$$\frac{\partial C_f}{\partial t} = \frac{D_F}{\left(p + (1-p)\rho_s \frac{\partial C_b}{\partial C_f}\right)} \frac{\partial^2 C_f}{\partial x^2} \text{ where } \frac{\partial C_b}{\partial C_f} \text{ is the binding capacity}$$

By positing the following :
- linear binding capacity
- constant porosity (no influence of time and salt concentration)
- no influence of salt concentration on diffusion coefficient
- proportionality between free chloride in solution and total amount of chloride

Fick's second law becomes : $\frac{\partial C_t}{\partial t} = D_a \frac{\partial^2 C_t}{\partial x^2}$ where D_a is the apparent diffusion coefficient, and then be solved. The solution of the equation will then be:

$$C_t(x,t) = C_s \left[1 - \text{erf}\left(\frac{x}{2\sqrt{D_a t}}\right)\right]$$

where C_s is the surface concentration.

By fitting this solution with chloride profiles concentration, the apparent diffusion coefficient can be determined. However, we have to take into account the fact that in our laboratory tests, the surface concentration is not constant in relation to time exposure.

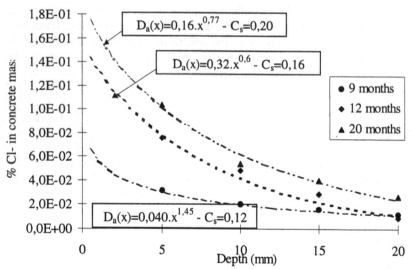

Fig. 9. Experimental chloride concentration results compared with Fick's second law curves

Fig. 9 shows an example of D_a calculation for the HSC samples. $C_s(t)$ is evaluated by assuming a linear relation between $C_s(t)$, $C_t(5 \text{ mm}, t)$ and $C_t(10 \text{ mm}, t)$.

Fig. 10 shows the experimental chloride profiles obtained on HSC samples after 6, 9, 12 and 20 months of exposure to the chloride environment. These results show that the unloaded sample will always have a lower chloride content when compare with the loaded samples. This concurs with the microscopic study of these samples indicating an increase in the microcracking density with the load applied.

Experimental results follow the same direction for ordinary concrete and very high strength concrete. As a result, it is obvious that, whatever the type of concrete, the microcrack network due to loading leads to an increase in chloride penetration. Similar results have already been obtained by François and Maso [1] on reinforced concrete beams made with ordinary concrete and Konin et al, [3].

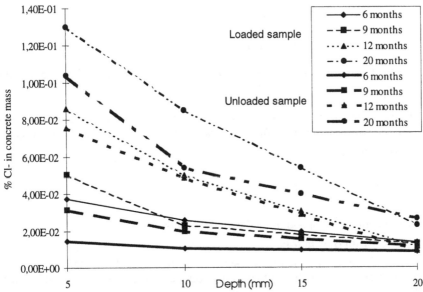

Fig. 10. Experimental chloride profiles for both HSC loaded sample and unloaded sample after 6, 9, 12 and 20 months of time exposure.

The problem was how to model the increase in chloride diffusivity in relation to the increase in microcracking density. At present, it remains impossible to establish a relationship between the load applied and the resultant microcracks. We thus decided to use the load level as the parameter of chloride increase. Because the tensile stress σ_s in the reinforcing bars is already used in civil engineering rule books as a durability parameter to check cracking [8], we decided to use it as the main parameter for a load function $L(\sigma_s)$. Although recent studies have indicated that the crack width has no influence on the corrosion process [9], σ_s can become a durability criterion as a threshold damage.

By correlating the experimental chloride profile with Fick's second law curve, we calculate the apparent diffusion coefficient of OC, HSC and VHSC of both loaded and unloaded samples at 9 months and 20 months exposure in a chloride environment. Results are presented in Table 1.

Table 1. Apparent diffusion coefficient in relation to the tensile stress of the re-bar

	σ_s (MPa)	0	172	179	186	200	215
$D_a(\sigma_s)$ cm^2/s	OC	1.23	1.56	1.63			
(9 months)	HSC	0.42		0.7	0.58		
	VHSC	0.29				0.43	0.47
$D_a(\sigma_s)$	OC	0.88	2.02	2.1			
(20 months)	HSC	0.58		1.45	1.6		
	VHSC	0.66				2	2.25

The apparent diffusion coefficient of the materials in unloaded state is called D_a^0.

To find the load function, the ratio $\dfrac{D_a}{D_a^0}$ is plotted in relation to the tensile stress of the re-bar (Fig. 11).

To take into account the damage increase in relation to time exposure, an aging function is updating the loading function was introduced [10]. Thus the relation between $D_a(\sigma_s)$ and D_a^0 can be written as follows:

$$D_a(\sigma_s,t) = D_a^0 \left[1 + L(\sigma_s)[1 + V(t)]\right]$$

The load function corresponding to the model on Fig. 11 is:

$L(\sigma_s) = 6E^{-8}\sigma_s^3$ with σ_s is in MPa

The origin of the aging function is taken conventionally at 9 months, the value of the aging function at 20 months is obtained by correlating the experimental points with the model.

For the time being, we do not propose an analytical equation for the aging function because this function takes into account different phenomena such as the degradation of concrete cover due to the corrosion of reinforcement, the microstructural evolution of concrete under different environmental actions (dry-humid cycles) and mechanical actions, etc..

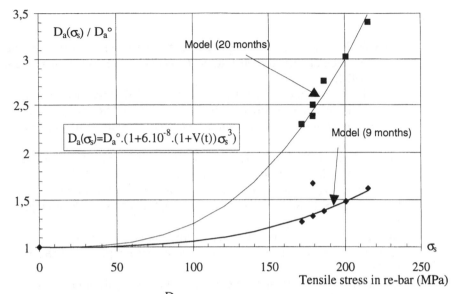

Fig. 11. Change in the ratio $\dfrac{D_a}{D_a^0}$ in relation to loading level and exposure time to the chloride environment.

The correlation proposed between loading level, aging properties of concrete subject to environmental actions and apparent diffusion coefficient represents a first and practical approach to characterize the influence of microcracks on the diffusion process.

To improve this approach, much work remains to be done such as:

• finding a direct correlation between concrete damage and loading level
• identifying and modeling the different parameters to introduce in the aging function
• improving the calculation of the apparent diffusion coefficient

5 Conclusion

Microcrack loaded samples present a strong degradation in their ability to withstand chloride penetration. Although other studies have shown that microcracks did not induce a change in chloride diffusivity [11], this can be explained through this study whose original feature is to evaluate diffusivity in a loaded state, that is with still open-microcracks due to the tensile loading state.

Because of the difficulties inherent in characterizing the microcrack network due to loading, the increase in the apparent diffusion coefficient is related to the loaded level by using the tensile stress in reinforcing bars as a main parameter [12].

This relationship could be used as a guideline to estimate the ingress of chloride into concrete subject to a tensile stress field due to the reinforcement.

6 References

1.   François, R. and Maso, J.C. (1988) Effect of damage in reinforced concrete on carbonation or chloride penetration', *Cement and Concrete Research* **18**, pp. 961-970.

2.   François, R., Arliguie, G. (1994). Effect of the ITZ damage on durability of reinforced concrete in chloride environment, *In Proc. MRS Symposium Bonding and Interfaces in Cementitous Materials*, volume 370, pp. 465-470.

3.   Konin, A., François, R. and Arliguie, G. (1995). Influence of the service load on chloride penetration into reinforced HSC, *In Proc. MRS Symposium Mechanisms of Chemical Degradation of Cement-based Systems,* (ed. K.L. Scrivener and J.F. Young), E & FN Spon, Boston, pp. 167-176.

4.   Ollivier, J.P. (1988). A non destructive procedure to observe the microcracks of concrete by scanning electron microscope' *C.C.R* **18**, pp. 35-43.

5.   Konin, A., François, R. and Arliguie, G. (1998). Analysis of progressive damage of reinforced ordinary and high performance concrete, *Materials and Structures*, Vol.31, January-February, pp.27-35.

6.   Yssorche, M.P. (1995). Microfissuration et Durabilité des BHP. PhD thesis, INSA, Toulouse, France.

7.   Nilsson, L.O., Poulsen, E., Sandberg, P., Sörensen and Klinghoffer, O. (1996). Chloride penetration into concrete, State of the art report. HETEK, report n°53.

8.   BAEL (1991).Règles de calcul du béton armé aux états limites

9.   François, R., Arliguie, G. and Maso, J.C. (1994). Durabilité du béton armé soumis à l'action des chlorures', *Annales de l'ITBTP* **529**, pp. 1-48.

10.   Gérard, B., Didry, O., Marchand, J., Breysse, D. and Hornain, H. (1995). Modelling the long-term durability of concrete for radioactive waste disposals, *In Proc. MRS Symposium Mechanisms of Chemical Degradation of Cement-based Systems,* (ed. K.L. Scrivener and J.F. Young), E & FN Spon, Boston, pp. 331-340.

11.   Locoge, P., Massat, M., Ollivier, J-P., Richet, C. (1992). Ion diffusion in microcracked concrete, Cement and Concrete Research, Vol. 22, pp. 431-438.

12.   Konin, A., François, R., Arliguie, G. (1998). Penetration of chlorides in relation to the microcracking state into reinforced ordinary and high strength concrete, to be published in *Materials and Structures*.

9   INFLUENCE OF THE INTERFACIAL TRANSITION ZONE ON THE RESISTANCE OF MORTAR TO CALCIUM LEACHING

A. DELAGRAVE, J. MARCHAND and M. PIGEON
Concrete Canada and Centre de Recherche Interuniversitaire sur le Béton
(CRIB), Université Laval, Québec, Canada

Abstract
The influence of the interfacial transition zone (ITZ) on the resistance of well-cured ordinary and high strength mortars (W/B=0.25 and 0.45) to calcium leaching was studied by soaking small disks of mortars in two different pH-controlled solutions over a six month period. The results indicate that the kinetics of degradation (calcium leaching) depends on the water/binder ratio, the type of binder, and the nature of the aggressive solution, but does not appear to be significantly influenced by the presence of numerous ITZ in the material. SEM observations revealed a clear front of degradation with no preferential decalcification around aggregates beyond this front.
Keywords: calcium leaching, chloride ions, ITZ, silica fume, water/binder ratio

1   Introduction

The interfacial transition zone (ITZ) formed at the vicinity of aggregate particles in concrete is a very thin layer of hydrated cement paste where the microstructure differs significantly from that of the bulk cement paste. Many studies indicate that this zone has a higher porosity and portlandite content [1, 2, 3]. From a theoretical point of view, the higher porosity of the ITZ, especially if they are interconnected, should facilitate the ingress of external aggressive agents as well as the leaching of calcium. Furthermore, the higher portlandite concentration in the ITZ should increase the potential risk of calcium leaching caused by the dissolution of $Ca(OH)_2$, and therefore detrimentally affect the resistance of concrete to ion penetration and modify the mechanical properties of cementitious systems.

The Interfacial Transition Zone in Cementitious Composites, edited by A. Katz, A. Bentur,
M. Alexander and G. Arliguie. Published in 1998 by E & FN Spon, 11 New Fetter Lane,
London EC4P 4EE, UK, ISBN: 0 419 24310 0

Many studies were conducted to investigate the microstructure of the ITZ and the influence of ITZ on mechanical properties. However, data on the influence of ITZ on the durability of cement-based materials remain limited. Recent results obtained by Bourdette [4] tend to indicate that the influence of ITZ on the resistance of mortars to chemical attack is limited. This paper presents the results of an investigation carried out to better understand the influence of ITZ on the durability of cement-based materials in contact with aggressive solutions.

2   Test Program

Four series of mixtures were cast using two water/binder ratios and four different binders. For each series, two mixtures of variable sand volume fractions were prepared. Disks 70 mm in diameter and 8 mm in thickness were soaked in 2 different pH-controlled solutions (with or without sodium chloride) for a six month period.

Mercury intrusion porosimetry, scanning electron microscopy, electron microprobe analyses, TGA/DTA, and X-ray fluorescence were the techniques used to characterize the evolution of the microstructural characteristics of the various mixtures over time. A sufficient number of disks of each mixture were soaked in the solutions to allow measurements to be made after 3 and 6 months of exposure.

3   Materials and Mixture Characteristics

The water/binder ratio of the first two series of mixtures was fixed at 0.25 while the water/binder ratio of the two other series was fixed at 0.45. The type of binder used in the preparation of the two high performance series (W/B=0.25) was an ASTM type III cement, and silica fume was used as partial cement replacement (6%) in one of these two series. The 0.45 water/binder ratio series were made with an ASTM type I cement, and silica fume was also used as partial cement replacement (6%) in one of the two series. The chemical analyses of the cements are given in Table 1.

For each of the four series, two mixtures with different sand volume fractions (0% and 50%) were prepared. A standardized crushed siliceous sand (Ottawa sand C-109) having a density of 2.60 was used. For the 0.25 water/binder ratio mixtures (with and without silica fume), a melamine-based superplasticizer was used at a dosage of 2.1% of dry materials by mass of cement. The composition of all mixtures is summarized in Table 2.

The complete results for the mortar mixtures with a sand volume content of 0% (neat pastes) are presented elsewhere [5]. Some of the results concerning these mixtures are presented again in this paper to facilitate the comparison between the neat pastes and the mortars, and to better visualize the influence of the interfacial transition zones on the durability of mortars in contact with aggressive solutions.

Table 1 - Chemical and mineralogical compositions of the cements

Chemical analysis (%)	ASTM III	ASTM I		ASTM III	ASTM I
SiO_2	20.48	20.09	Bogue Composition		
$Al2O_3$	4.03	3.87	C_3S	68.7	68.7
Fe_2O_3	1.78	1.69	C_2S	6.9	5.8
CaO	64.73	63.82	C_3A	7.7	7.4
SO_3	3.33	3.50	C_4AF	5.4	5.1
MgO	2.31	2.22	Blaine (cm^2/g)	5351	4616
Na_2O	0.36	0.30			
K_2O	0.34	0.39			
TiO_2	0.17	0.16			
MnO	0.05	0.05			

Table 2 - Mixture compositions

Mixture	W/B	Cement	Silica Fume (%)	Sand volume (%)
M25-0	0.25	ASTM III	0	0
M25SF-0	0.25	ASTM III	6	0
M45-0	0.45	ASTM I	0	0
M45SF-0	0.45	ASTM I	6	0
M25-50	0.25	ASTM III	0	50
M25SF-50	0.25	ASTM III	6	50
M45-50	0.45	ASTM I	0	50
M45SF-50	0.45	ASTM I	6	50

4 Experimental Procedures

All specimens were cast in plastic molds (diameter = 70 mm, height = 200 mm). The molds were sealed and rotated for the first 24 hours to prevent bleeding and segregation. At the end of this period, the specimens were demolded and immersed (at room temperature) in a saturated lime solution for a 2-month period.

At the end of the curing period, disks 8 mm in thickness were cut from the cylinders and immersed in two different aggressive solutions. A number of disks were set aside for the determination of the initial characteristics of the mixtures. One 50-litre plastic tank was filled with distilled water and another contained a 3% NaCl (by mass) solution. The initial pH level of the solutions was 7. The pH level was adjusted manually every 24 hours by adding HCl (1N) to the solution already containing chloride ions and HNO_3 (0.1N) to the chloride free solution.

After 3 and 6 months of exposure, two disks of each mixture were removed from the solutions and immediately vacuum dried for a minimum period of 10 days to prevent any further chemical reactions before the different analyses were performed, except for the samples required for the mercury intrusion porosimetry measurements. These samples were immersed in propan-2-ol for a minimum period of 21 days according to the procedure described by Feldman and Beaudoin [6]. This sample preparation technique was used to reduce as much as possible the pore structure alteration due to drying. After the immersion in propan-2-ol, all samples (representing the full thickness of the disks) were vacuum dried for 24 hours prior to testing. The minimum intruding pressure was 2.6 kPa and the maximum pressure was 207 MPa. The contact angle assumed for all samples was 130°.

For the scanning electron microscope and the microprobe analyses, the disks were broken into small parts to expose the total internal surface (8 mm thick). They were then impregnated with an epoxy resin, polished, and coated with carbon. The microprobe measurements were performed along imaginary lines extending 4000 µm from the external surface in contact with the aggressive solution towards the internal part of the disks. The concentration of calcium and silicium were determined along these lines. For the mixture containing sand (50% sand volume content), the microprobe measurements were performed in the paste fraction of the mortars.

The chemical and mineralogical compositions of all mixtures were determined at the end of the curing period and after 6 months of exposure to the aggressive solutions. In all cases, these chemical compositions were determined by TGA/DTA and X-ray fluorescence analyses. All analyses were performed on powdered samples (passing a 75 mm sieve) representing the full thickness of the disks.

5   Test Results

5.1 SEM Observations

Figure 1 presents the microstructure of the M45-50 mortar mixture at the end of the curing period (control specimen). As can be seen on the Figure, the aggregates are surrounded by a white ring of approximately 10 to 20 µm in thickness. The energy dispersion measurements revealed that calcium was the most important constituent of this zone.

The microstructure of M45(SF)-50 was found to be quite similar to that of M45-50 except that the thickness of the calcium shell appeared to be limited to approximately 10 µm. For M25-50 and M25(SF)-50, however, the difference between the microstructure at the vicinity of the aggregates and that of the bulk cement paste was much less visible and it was not possible to identify a calcium rich zone near the aggregates (see Figure 2). Furthermore, in these mixtures, even after two months of curing, numerous anhydrous grains were observed. This was confirmed by the TGA/DTA analyses.

Figure 3 presents the microstructure of the M45-50 mortar mixture after 3 months of exposure to the solution maintained at a pH level of 7 without chloride ions. As can be seen

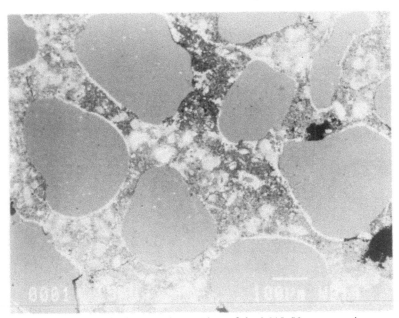

Figure 1 - Typical SEM observation of the M45-50 mortar mixture
at the end of the curing period

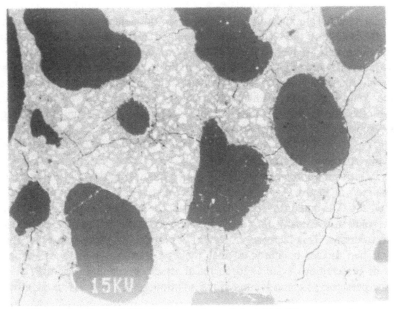

Figure 2 - Typical SEM observation of the M25-50 mortar mixture
at the end of the curing period

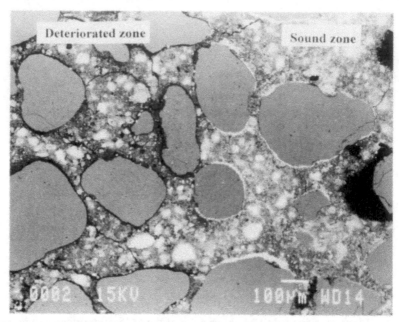

Figure 3 - Dissolution front observed on M45-50 after 3 months of exposure
to the chloride free solution

on the Figure, the dissolution front is very clear. In the deteriorated zone (in contact with
the aggressive solution), there is no more calcium left at the vicinity of the aggregates. The
SEM observations also indicated that the bulk paste in this zone was more porous than in
the internal (sound) part of the disk. In the sound part, there was no sign of degradation in
the interfacial transition zones and the microstructure of the bulk paste appeared similar to
that of the control specimen. Similar observations were made for all mortar mixtures in
contact with the aggressive solutions, although the degradation around the aggregates was
much less visible for M25-50 and M25(FS)-50.

5.2 Mercury Intrusion Porosimetry
The mercury intrusion porosimetry test results for the neat pastes and the mortar mixtures
are summarized in Figure 4. As could be expected, the 0.25 mixtures are always less
porous than the 0.45 mixtures, irrespective of the time of exposure or the nature of the
aggressive solution. Generally, the total porosity increases with the time of exposure to the
aggressive solutions. This increase, however, appears to be more important for the 0.45
mixtures. It can further be observed that the use of silica fume tends to improve the
resistance of cementitious systems to chemical attack, particularly for the 0.45 mixtures,
and that the presence of chloride ions in the solution generally tends to slightly accelerate

the increase in porosity. These observations apply to all mixtures, irrespective of the sand volume fraction.

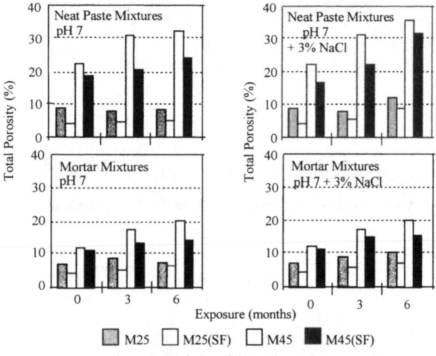

Figure 4 - Evolution of the total porosity

5.3 Microprobe analyses

Figure 5 presents the microprobe profiles after 6 months of exposure to the aggressive solutions for all mortar mixtures. In order to facilitate the comparison between mortars and neat pastes, the calcium profiles for the neat pastes are superimposed on the diagrams. The straight line on each diagram represents approximately the minimum concentration of calcium in the control specimens (i.e. in the C-S-H). The concentrations above this line correspond to the calcium in portlandite crystals or in anhydrous silicates, and the concentrations below, to the calcium in deteriorated cement paste including C-S-H.

As can be seen on Figure 5, the silicium concentration appears to be constant over the full thickness of the disk, irrespective the nature of the aggressive solution or the time of exposure. The situation, however, is quite different for the calcium profiles. Near the external part of the disks, the calcium concentration is always lower than in the internal part. This decalcified zone appears to be more important for the 0.45 mixtures. For example, the depth of decalcification after 6 months of exposure to the chloride free solution is approximately 1000 µm for the 0.25 mixtures while for the 0.45 mixtures, there

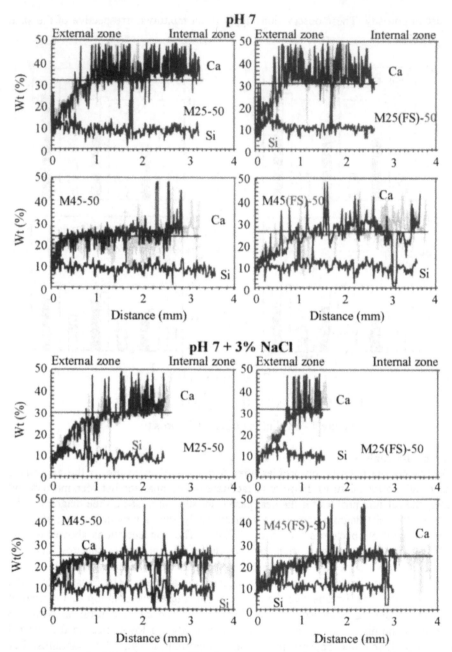

Figure 5 - Microprobe profiles for the mortar and neat paste samples
after 6 months of exposure to the aggressive solutions

very significant decalcification thoughout. Figure 5 also shows that the use of silica fume ontributes to slow down the decalcification process, and that decalcification is always more important for the mixtures immersed in the chloride solution.

It is very interesting to note on Figure 5 that, irrespective of the water/binder ratio, the type of binder or the nature of the aggressive solution, the calcium profiles for the neat pastes are quite similar to those in the corresponding mortars.

.4 Chemical Analyses

Results of the TGA/DTA and X-ray fluorescence analyses are presented in Table 3. For the mortar mixtures, the results of the different analyses were recalculated on a unit volume of paste basis. As can be seen in the Table, the results clearly underline that the use of silica fume, as well as a reduction of the water/binder ratio, contributes to improve the resistance of cementitious materials to calcium leaching. There is always more calcium and portlandite left in the 0.25 mixtures and in the mixtures containing silica fume. These results also confirm that the presence of chloride ions in the aggressive solution contributes to accelerate calcium leaching.

It is again very interesting to note (in Table 3) the similarity of results for the neat pastes and the mortars (except perhaps for the 0.45 mixtures for which the mortars appear to be less deteriorated than the neat pastes).

Table 3 - TGA/DTA and X-Ray Fluorescence Analyses after 6 months of exposure (percentage of initial value)

Mixture	pH 7 Ca(OH)$_2$	pH 7 + 3% NaCl Ca(OH)$_2$	pH 7 CaO	pH 7 + 3% NaCl CaO
M25-0	80	62	80	77
M25SF-0	83	95	90	87
M45-0	0	0	59	47
M45SF-0	54	0	82	73
M25-50	80	66	97	87
M25SF-50	83	95	92	90
M45-50	41	0	73	59
M45SF-50	66	0	86	75

Discussion

For all mixtures tested (both pastes and mortars), immersion in the two aggressive solutions used in these series of experiments was found to cause the leaching of calcium from both the Ca(OH)2 and C-S-H phases, as it is particularly clear from the microprobe profiles and the TGA/DTA and X-ray fluorescence analyses. The interfacial transition zone

(ITZ) was observed to be more susceptible to calcium leaching than the bulk cement paste. In the deteriorated part of the specimens, there was often no more calcium left in the ITZ while there was still some calcium left in the bulk paste. However, the dissolution front of calcium was found to be very neat, and, beyond this front, no clear signs of decalcification around aggregates were seen. These results are in good agreement with those published by Bourdette [4].

Even if the hydrates in the ITZ are more soluble than those in the bulk cement paste, the microprope profiles and the TGA/DTA analyses clearly show that the kinetics of degradation of the mortars tested are similar to those of the neat pastes. An extensive description of the degradation mechanisms is beyond the scope of this paper and can be found elsewhere [7, 8, 9]. However, it must be remembered that the hydrates are in thermodynamic equilibrium with the pore solution and that small variations in the composition of this solution readily modifies the equilibrium and lead to the dissolution of certain hydrated phases. Adenot [8] showed that the degradation of cementitious systems is made of several dissolution fronts depending on the relative solubility of the hydrates. The rate at which these fronts evolve depends not only on the solubility of the hydrates but mainly on the diffusivity of calcium. The fact that the mortar mixtures underwent decalcification at the same rate as neat cement pastes clearly indicate that, even if the ITZ are more susceptible to calcium leaching, the overall phenomenon is governed by the bulk cement paste which has the lowest diffusivity. The kinetics of degradation are thus not significantly influenced by the presence of numerous ITZ in the material. Of course, the fact that the ITZ do not significantly modify the calcium leaching mechanisms do not necessarily imply that they have no influence on other properties of the material, like mechanical properties for example [10].

7   Conclusion

The test results presented in this paper indicate that the ITZ do not significantly influence the resistance of mortar to calcium leaching. They also confirm the beneficial influence of low water/binder ratios and of the use of silica fume to improve the resistance of cement based materials to calcium leaching. The detrimental effect of chloride ions was also observed.

8   Acknowledgements

The authors are grateful to the Natural Sciences and Engineering Research Council of Canada for its financial support for this project which is part of the research program of the Network of Centers of Excellence on High Performance Concrete (Concrete Canada). The authors also whish to thank Hélène Desrosiers for her fine help in performing the laboratory experiments.

References

[1]   Maso, J.C. (1980) La liaison entre les granulats et la pâte de ciment hydraté, *7th International Congress on the Chemistry of Cement*, Paris, Principal Report, I, 7-I, 3-14.

[2]   Scrivener, K.L., Bentur, A., Pratt, P.L.(1988) Quantitative characterization of the transition zone in high strength concretes, *Advances in Cement Research*, 1, 230-237.

[3]   Bourdette, B., Ringot, E., Ollivier, J.P. (1995) Modelling of the transition zone porosity, *Cement and Concrete Research*, 25, 4, 741-751.

[4]   Bourdette, B. (1994) Durabilité des mortiers, Ph.D. Thesis, INSA-UPS Toulouse, France.

[5]   Delagrave, A., Pigeon, M., Marchand, J. (1996) Durability of high performance cement pastes in contact with chloride solutions, Fourth International Symposium on the Utilization of High Strength/High Performance Concrete, Paris, 2, 479-488.

[6]   Feldman, R.F., Beaudoin, J.J. (1991) *Cement and Concrete Research*, 21, 297-308.

[7]   Revertégat, E., Richet, C., Gégout, P. (1992) Effect of pH on the durability of cement pastes, *Cement and Concrete Research*, 22, 2/3, 259-272.

[8]   Adenot, F. (1992) Durabilité du béton: caractérisation et modélisation des processus physiques et chimiques de dégradation du ciment, Ph.D. Thesis, Université d'Orléans, France.

[9]   Marchand, J., Gérard, B., Delagrave, A. (1997) Ion Transport Mechanisms in Cement-Based Materials and Composites, *In Materials Science of Concrete V*, American Ceramic Society.

[10]   Carde, C., François, R. (1997) Effect of ITZ Leaching on the Durability of Cement-Based Materials, *Cement and Concrete Research*, 27, 971-978.

10 EVALUATION OF THE INTERFACIAL TRANSITION ZONE IN CONCRETE AFFECTED BY ALKALI–SILICA REACTION

P. RIVARD and G. BALLIVY
Civil Engineering Department, University of Sherbrooke, Sherbrooke, Canada
B. FOURNIER
Canada Centre for Mineral and Energy Technology, Advanced Concrete Technology Program, Ottawa, Canada

Abstract
Alkali-silica reaction (ASR) has been observed for 63 weeks in accelerated conditions at different stages of its evolution on laboratory concrete samples made with reactive Potsdam sandstone. Samples were submitted to an automatic petrographic technique for quantification of damage related to ASR. The evolution of the interfacial transition zone (ITZ) has been deeply investigated with SEM and microprobe in order to get a qualitative analysis of alkali migration. Other petrographic observations have been carried out using the optical method of Damage Rating Index. Work showed that damage (cracking) caused by ASR to concrete incorporating aggregate such as Potsdam sandstone can be estimated by analysing the ITZ. It showed also that the transition zone is very porous, which allows alkali hydroxides to penetrate the aggregate particle. These results are compared with observations of samples taken from concrete dam subjected to ASR in Eastern Canada; the ages of these concretes are up to 70 years old, so it is possible to extrapolate the results of this experimental work.
Keywords : alkali-silica reaction, concrete damage, image analysis, microcracking, Potsdam sandstone, reaction rim, interfacial transition zone.

1 Introduction

Since the pioneering work of Stanton in the late 1940's [1], alkali-silica reaction (ASR) has been reported to affect concrete structures in more than 50 countries around the world. The above, which has often been qualified as the concrete cancer, consists of chemical reactions between dissolved substances in the concrete pore fluid (i.e. alkali hydroxides) and some siliceous mineral phases in the aggregate particles. These reactions result in the formation of a secondary product, a calcium-rich alkali-silica gel, which absorbs moisture, generates pressure on the cement paste, and eventually induce the cracking and distress of the affected concrete member.

During the on-going process of ASR, intense chemical/physical reactions occur in the interfacial zone between the cement paste and the reactive aggregate particles. Breton and Ballivy [2] reported that ITZ constitutes an area where ionic migrations are promoted. For an aggregate such as granite, the external periphery act as a impermeable membrane, stopping alkali migration into particle [3].

The Interfacial Transition Zone in Cementitious Composites, edited by A. Katz, A. Bentur, M. Alexander and G. Arliguie. Published in 1998 by E & FN Spon, 11 New Fetter Lane, London EC4P 4EE, UK, ISBN: 0 419 24310 0

Thaulow and Knudsen [4] studied the reaction of a 9 mm core of highly-reactive opaline material embedded in cement paste after eight months of moist curing. The authors identified the following four zones (from inside the opal towards the cement paste) :

1. Reacted opal with reduced Si and increased Ca contents,
2. Reacted opal with reduced Ca and increased Si contents, and traces of K,
3. Zone in the cement paste impregnated with calcium-rich alkali (K)-silica gel,
4. Bulk cement paste not impregnated by gel.

Detailed chemical micro-analysis showed the penetration of potassium from the cement paste pore solution through all of the zone 1 described above, and even to the reacted opal interface, although the amounts involved were only slightly above the detection level. Sodium was more abundant in the cement used in the above study but could hardly be chemically detected.

In a similar study, Baker and Poole [5] also observed alkali ions migrating into the opal aggregate from the cement paste. They suggested that calcium silica hydrates (CSH) normally formed during cement hydration may not have completely developed under high alkali concentration in the interfacial zone with opal, thus allowing alkali ions to attack opal and produce the reaction gel.

Davies and Oberholster [6] examined the interface between cement paste and quartzite aggregate particles. The authors observed that K_2O/SiO_2 and CaO/SiO_2 were stable at a ratio of about 0.2 all the way through the aggregate particle to the unreacted silica core. Following the classical model of ASR, they concluded that the alkali hydroxides must precede Ca ions into the reaction zone, although Ca content may eventually become higher.

2 Scope of work

The work presented in this paper deals with the evaluation of the development of damage in the interfacial transition zone (ITZ) between a reactive quartzitic sandstone aggregate and a high-alkali cement paste in concrete. A series of concrete prisms were made with the above aggregate according to CSA A23.2-14A Concrete Prism Test procedure. The prisms were stored in accelerated test conditions in the laboratory (38°C, R.H. > 95%), and their expansion monitored at regular intervals. Polished sections were then prepared from the above prisms at selected levels of expansion for a detailed petrographic examination. Similar sections were also prepared from cores collected from a large hydraulic dam made with the same aggregate and suffering from ASR. The sections were examined petrographically under a petrographic microscope connected to an image analyzing system in order to quantify the current condition of the concrete. The specific objectives of this investigation were:

- Monitor the evolution of the ITZ in the above concretes,
- Correlate the petrographic features thus observed at the ITZ with the amount of expansion obtained with the concrete specimens,
- Compare the above information to the current condition of concrete cores collected from a large hydraulic dam incorporating the same aggregate and suffering from ASR.

3 Materials

The coarse aggregate selected for this investigation is a quarzitic sandstone from the Potsdam Group. It has been used in a number of concrete structures in the Montreal area (Canada), which show various degrees of distress due to ASR [7,8,9]. The reactive phase in the aggregate is the siliceous material cementing quartz grains [8]. A characteristic feature of the reacted aggregate particle is a dark rim visible at the internal periphery of the particles. Non-reactive fine aggregate derived from granite was used with the sandstone aggregate for the preparation of the concrete prisms in the laboratory.

A normal ASTM Type I high-alkali cement was used in the preparation of the concrete. According to the CSA A23.2-14A test procedure, NaOH pellets were added to the mixture water to raise the alkali content of the system to 1.25% Na2O equivalent by cement mass.

4 Experiment

4.1 Preparation of Concrete Prisms in the Laboratory

A set of twelve concrete prisms, 75 by 75 by 300 mm in size, were made from a concrete mixture incorporating the Potsdam sandstone, a nominal cement content of 420 kg/m^3 and a water-to-cement ratio of 0.40. As mentioned before, NaOH pellets were added to raise the total alkali content of the system to 5.3 kg/m^3. After a curing period of 24 hours, test prisms were stored in plastic pails lined with damp terry cloth. The pails were then stored at 38°C and the test prisms measured periodically for the development of expansion and cracking due to ASR. At selected expansion levels, one prism was removed from the storage condition, and cut in two or three 50-mm thick sections in order to be analysed.

4.2 Concrete Core from the hydraulic dam

A 10-m long, 300-mm in diameter core was collected from a large gravity dam located in the Montreal area, Canada. This dam, constructed with Potsdam sandstone in three steps from 1929 to 1961, has a length of 865 m, and can impound a flow rate up to 10 500 m^3/s. A hydraulic head of 25 m permits a total capacity of 1645 MW of electricity to be generated from 36 turbines. This dam is suffering from ASR on most of its components [8]. Ballivy and *al.* [9] reported that pressure generated by silica gel is up to 4 MPa. The core was examined and locations showing various degrees of damage were selected for detailed petrographic examination.

4.3 Experimental procedure - Quantitative Petrographic Analysis

Each of the slices cut from the test prisms and concrete cores was first impregnated with a fluorescent epoxy. Detailed information on sample preparation is given in [10].

The quantitative petrographic analysis was performed on the polished sections under a stereobinocular microscope at 20x magnification. UV light was used for detection of cracks which have been impregnated by the UV loaded epoxy. Images were acquired and recorded using a high-sensibility camera. A program was developed for the analysis of the images and parameters such as total length of cracks, crack density and orientation were automatically calculated for each image [10]. Information collected from each image was then averaged to have a rating for the whole polished section.

The polished sections prepared from the test prisms and the cores were also examined to determine their Damage Rating Index (DRI). This quantitative procedure developed by Grattan-Bellew [11] consists in a visual examination of polished section under a stereobinocular microscope at 10x magnification. The result is a number, which is based on various defects (reaction rim, cracks, etc.) affecting concrete suffering of ASR. Some sections were also examined under a low-vacuum SEM and a microprobe to study the chemical changes occurring at the ITZ.

5 Results

5.1 Alkali mapping

A chemical mapping was carried out at the interfacial zone around about 30 sandstone particles from the test prisms and cores. No particular alkali concentration was observed around the particles at early ages. With the increasing reaction/expansion processes, K ions started to be detected concentrating close to the external periphery of the particles (Fig. 1 - expansion level of 0.036%). At an expansion level of 0.075%, a clear halo of potassium was observed around the reactive particles, while Ca and Na showed no detectable concentration at that location (Fig. 2). At a more advanced stage of expansion (i.e. 0.14%), K was found scattered inside the sandstone aggregate particles, but especially between the well-rounded quartz grains where a secondary alkali-silica gel was progressively replacing the original siliceous cement (Fig. 3). Reaction is almost finished at this expansion level [10]. Alkalies are conceivably not available anymore, so there is no more concentration process around particles. Alkali penetration is relatively fast, indicating that the transition zone and aggregate particle are very porous, which would promote ASR.

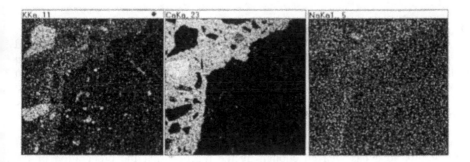

Fig.1 K, Ca and Na concentration at expansion level of 0.036%

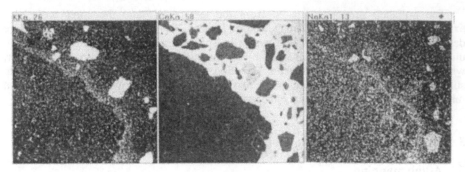

Fig.2 K, Ca and Na concentration at expansion level of 0.75%

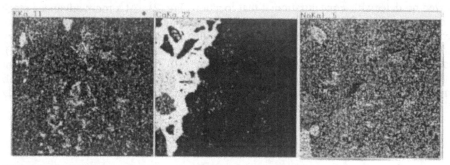

Fig.3 K, Ca and Na concentration at expansion level of 0.14%

5.2 Automatic Damage Quantification (Image Analysis)

The results of the most representative parameter, for instance crack density, are plotted on Fig. 4. It shows a reasonably good correlation between cracking and expansion level; however, high levels of expansion surprisingly developed without the formation of extensive measurable cracking. Fig. 5 shows a particle with a reaction rim. No significant damage is observed in this zone; the well-rounded quartz grains at the internal border of aggregate particles are still fairly well cemented. However, the internal portion of the particle is completely filled by fluorescent resin; this suggests that the intergranular siliceous cement has progressively disappeared by the process of ASR, or this zone has weaken enough by the reaction that the preparation of the polished section has created significant porosity that has been filled subsequently by the epoxy resin.

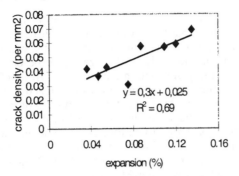

Fig.4 Automatic quantification of crack density

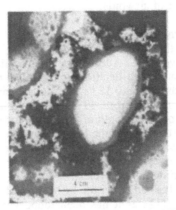

Fig.5 Reactive particle showing reaction rim at ITZ as seen in UV light

5.3 Damage Rating Index (DRI)

A total of five laboratory test samples were subjected to DRI procedure. The results, given on Fig. 6, generally show a good correlation between increasing DRI numbers and increasing expansion levels of the concrete prisms from which the section was prepared.

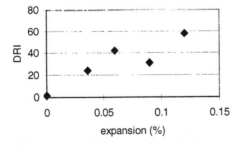

Fig.6 Damage Rating Index of some sample vs expansion level

No sign of reaction was observed on the sample at early age (expansion level of 0%). However, as the expansion reaches higher levels, a larger number of reaction rims were observed affecting numerous aggregate particles. At 0.036%, some particles started to show reaction rims developing, but these were generally difficult to identify. At that stage, no particular damage was observed in the cement paste, while a few cracks were observed in the aggregate particles only. Silica gel could be observed in some very fine cracks and between quartz grains within the particles. At 0.06%, a larger quantity of reactive particles were showing more or less distinct reaction rims; the rims were about 0.25 mm in thickness on an average. Larger amounts of gel were found in the cement paste. Some gel deposits were also observed in the air voids of the cement paste. At 0.09%, reaction rims started to be much more evident and easier to identify; the damage seemed to be relatively similar to the specimen at the 0.06%, although more gel was observed in voids of the cement paste. At 0.12 and 0.14%, reaction rims are much more well defined, easily countable and affect a large number of aggregate particles. Cracks were observed in larger amounts within the particles while gel can be observed in a large number of air voids.

5.4 Results for Beauharnois Dam
A number of polished sections were examined from different locations along the 10-m core collected in the above dam. Automatic quantification, as well as DRI, showed high values, which indicates high deterioration of the structure. According to these data, expansion could be as high as 0.12 %, since crack density values varied from 0.08 to 0.14, and DRI values ranged from 40 to 110. Variations indicate that concrete condition varies widely along the length of the core. The highest values were obtained close to the surface of the dam, while the lowest values were obtained from the middle part (from 4 to 6 m). Visual deterioration is much more significant in the first few centimetres of the structure due to the accelerating effect of the environment (freezing/thawing and wetting/drying cycles).

Chemical mapping performed under the SEM showed that there is no concentration of alkalies (K, Na) in the ITZ, as it has been found with high expansion laboratory concrete. Also, a large proportion of the intergranular siliceous cement within the sandstone particles has been replaced by alkali-silica gel. Microprobe line scan of the ITZ reveals that no chemical changes occurred in this zone regarding alkalies (K, Na, Ca).

Petrographic features observed on Beauharnois samples are not very different from test prisms, even if these were submitted to accelerated conditions. In both cases, reaction rims were found around every reactive particles and intergranular siliceous cement is dissolved inside these particles. Cracks are wider with Beauharnois samples, which seems normal since they come from a much bigger volume of concrete with very coarse aggregates (diam. 5-7 cm).

6 Concluding remarks

The Potsdam sandstone is a porous rock, and so the interface rock/cement paste promotes alkali-silica reaction, instead of acting as a impermeable membrane. This zone allows alkali hydroxides to penetrate the aggregate particle.

With Potsdam sandstone, or similar rock type aggregate, damage (cracking) caused by ASR to concrete can be estimated by analysing the interfacial transition zone. Chemical mapping showed that there is a migration of alkali from cement paste toward reactive aggregates. Potassium seems to be the only reliable marker of the reaction. At expansion levels between 0,04 % and 0,12 %, K is present in relatively high concentration at the ITZ. Then, it moves inside the aggregate, and no specific concentration can be seen at the ITZ.

Petrographic analysis showed that development of reaction rim at the ITZ is correlated to concrete expansion, as well as damage caused to concrete. The dark rim seems to be related to an optical phenomenon, since chemical mapping and microprobe line scan did not show any difference between this zone and the internal zone. Rims appear clearly at expansion of about 0,06%, as some cracks begin to be found in cement paste.

Acknowledgements

The authors would like to thank Hydro-Quebec, CANMET, and the National Sciences and Engineering Research Council of Canada (NSERC) for their financial support. Dr. R. Lastra, from CANMET, is acknowledged for his precious help with the image analysis program. We also thank Dr. P. Grattan-Bellew of the National Research Council of Canada for allowing us to use his equipment for the Damage Rating Index.

References

1. STANTON, T.E. (1960) Expansion of concrete through reaction between cement and aggregate, *Proc. American Society of Civ. Engrs*, Vol. 66, p. 1781-1811.
2. BRETON, D., BALLIVY, G. (1996) Correlation between the transport properties of the transition zone and its mineral composition and microstructure, *The modelling of microstructure and its potential for studying transport and durability*, NATO edition, p.361-372.
3. BÉRARD, J., ROUX, R. (1986) La viabilité des bétons du Québec : le rôle des granulats, *Canadian Journal of Civil Engineering*, vol. 13, p.12-24.
4. THAULOW, N., KNUDSEN, T. (1975) Quantitative microanalysis of the reaction zone between cement paste and opal, *Proceeding of the Symposium on alkali-aggregate reactions, preventive measures*, Reykjavik (Iceland), August 1975, p.189-203.
5. BAKER, A.F., POOLE, A.B. (1980) Cement hydrate development at opal-cement interfaces and alkali-silica reactivity, *Q. J. Eng. Geol. London*, Vol. 13, p.249-254.
6. DAVIES, G., OBERHOLSTER, R.E. (1986) Chemical and swell properties of the alkali-silica reaction product, *Proceedings of the 8th International congress on the chemistry of cement*, Rio de Janeiro (Brazil), Vol. V, p. 249-255.
7. BÉRARD, J., LAPIERRE, N. (1977) Réactivités aux alcalis du grès de Potsdam dans les bétons, *Canadian Journal of Civil Engineering*, Vol. 4, No. 3, p.332-344.
8. ABET, M., Y. (1984) Recherche sur les causes des déformations de l'aménagement hydro-électrique de Beauharnois, Master Thesis, École Polytechnique de Montréal, (Canada).
9. BALLIVY, G., BOIS, A.P., SALEH, K., RIVEST, M. (1995) Monitoring of the stresses induced by AAR in the Beauharnois concrete gravity dam, Proceedings 2nd International Conference on AAR, USCOLD, Chattanooga (USA), October, p.343-358
10. RIVARD, P. (1998) Quantification de l'endommagement du béton atteint de réaction alcalis-silice par analyse d'images, Master Thesis, University of Sherbrooke, Sherbrooke (Canada).
11. GRATTAN-BELLEW, P.E., DANAY, A. (1992) Comparaison of laboratory and field evaluation of AAR in large dams, Proceedings *International Conference on Concrete AAR in Hydroelectric Plants and Dams*, Sept. 28 to Oct 2. 1992, Canadian Electrical Association in association with Canadian National Committee of the International Commission on Large Dams, p.23.

PART FOUR
ROLE OF ITZ IN CONTROLLING DURABILITY OF CONCRETE

11 INFLUENCE OF THE INTERFACIAL TRANSITION ZONE ON THE FROST RESISTANCE OF CALCAREOUS CONCRETES

F. RENAUD-CASBONNE, M. BELLANGER and F. HOMAND
LEGO Nancy, Ecole Nationale Supérieure de Géologie, Nancy, France

Abstract
The general framework of this work is the study of the behavior of concretes containing limestone aggregates of Lorraine (north eastern of France) submitted to freezing and thawing cycles. Properties of these concretes are compared to those of a siliceous concrete. Measurements of ultrasonic waves velocities, gas permeability and compressive strength allow to put in obviousness that the degradations occur in majority before 25 freezing and thawing cycles and evolve then few between 25 and 300 cycles. SEM observations show the appearance of secondary ettringite in relationship with the creation and the opening of cracks. Results indicate that the homogeneity of paste-aggregate contact influences frost resistance, that is weaker for concretes with siliceous aggregates than for concretes with calcareous aggregates.
Keywords: calcareous aggregates, durability, freezing and thawing cycles, frost resistance, paste-aggregate contact.

1 Introduction

The aim of this study is to put in obviousness the mechanisms occurring during freezing and thawing cycles in the case of calcareous aggregates, reactive with cement paste. In Lorraine, exploitation of calcareous deposits is in full expansion. Indeed, their abundance and their weak cost confer them a great interest because siliceous alluvium rarefy. However, the utilization of limestones has been braked until present, because they are often classified out of specifications, held account of their frost behavior and the climate of Lorraine. They present nevertheless a great reactivity with cement demonstrating by the appearance of a compact transition zone between cement paste and calcareous aggregates [1].
Numerous authors have studied the behavior of concretes during freezing and thawing cycles. All theories explain degradations of concrete by movement of water during ice formation giving rise important pressures causing cracks when pressures become superior to the traction resistance of paste. Specifications recommend to realize compact concretes, a W/C ratio minima and a cement content superior to a specific value, the use of frost resistant aggregates and especially the creation of a network containing entrained air which serve as expansion chamber. The realized studies until now consider only inert aggregates.
A previous study has shown that concretes made with limestones of Lorraine have good mechanical performances because of the great continuity between paste and

The Interfacial Transition Zone in Cementitious Composites, edited by A. Katz, A. Bentur, M. Alexander and G. Arliguie. Published in 1998 by E & FN Spon, 11 New Fetter Lane, London EC4P 4EE, UK, ISBN: 0 419 24310 0

aggregates and despite the weak qualities of limestone aggregates [2]. Therefore it appears interesting to study the mechanical and the physical behaviors to freezing and thawing cycles of these concretes and to analyze possible modifications of the hydrates constituting cement paste.

2 Materials and composition of the concretes

2.1 Constituents of the concretes

2.1.1 Aggregates

Three calcareous and one siliceous aggregates are used.
1. *Limestone of Jaumont*: superior Bajocian, a weak, oolithic and bioclastic rock.
2. *Limestone of Attigneville*: superior Bajocian, a hard and sublithographic rock.
3. *Limestone of Gudmont*: superior Oxfordian, a mixture of a sublithographic and an oolithic limestones hard rock.
4. *Siliceous aggregate*: fluviatile crushed alluvium of the Moselle.

2.1.2 Cement

The cement is a CPJ-CEM II/A 32,5R, containing 57% of C3S, 14% C2S, 10% C3A, 10% C4AF and 4,5% gypsum.

2.2 Composition of the concretes

The compositions of the concretes are presented in the table 1. A W/C ratio equal to 0.5 is retained. This ratio characterizes the required water for cement hydration, but it does not take in consideration the possibility of water absorption by the porous calcareous aggregates [3]. An additional water volume corresponding to the water absorbed by the aggregates during the batch, is added. In order to determine this volume an aggregate sample is weighed and immersed in water during 10 minutes. The surface is dried and the sample is again weighed. The difference of masses corresponds to the water absorption.

A 0/5 sand and a 5/10 gravel are used in the following proportions: 45% of sand and 55% of gravel. Concretes are deliberately realized without entrained air because the aim is to study modifications of structures after freezing and thawing cycles, and not to optimize formulations.

The consistency of concretes is determined by measurements to the Abrams cone. It varies from very plastic to fluid, with respect to the nature of the aggregate. The air content is determined by measurements with an aerometer: it is about of 1%. The concretes are cured during 28 days in water maintained at the standard temperature of 20°C.

Table 1 - Composition of the concretes (kg/m^3)

	Aggregate			
	Gudmont limestone (G)	Jaumont limestone (J)	Attigneville limestone (A)	Siliceous (S)
gravel 5/10	821	750	775	892
sand 0/5	665	597	635	713
cement	483	483	483	483
hydration water	241	241	241	241
absorbed water	27	60	31	11

3 Experimental methods

After a cure of 28 days, the concretes are submitted to freezing and thawing cycles of 12 hours with a rate of frost of 8°C/hour and temperatures varying from +5°C to -15°C with landings of frost of 3h and thaw of 4h.
Before the experimental tests the samples are completely saturated by immersion in water under void after drying. They are placed in sealed plastic sleeves during freezing and thawing cycles to maintain constant the saturation.
Different tests are realized after 5, 25 and 300 cycles. The mechanical properties are characterized by the compressive strength. The physical properties are characterized by ultrasonic waves velocities, which is influenced by the crack density. The measurements of gas permeability give indications on the degree of contact of the porous system elements and therefore on the possible cracks opening. To these tests are added SEM observations to put in obviousness the possible modification of hydrates.

4 Results

4.1 SEM Observations

4.1.1 Reference Samples

1. *Siliceous concrete*: the paste is constituted mainly by CSH gel, portlandite and ettringite. It is relatively loose because of the presence of primary ettringite crystallized under form of sticks in the cavities of the paste. It exists a discontinuity between paste and aggregates (photo 1). The transition zone with a high porosity is composed of hydrates of great dimensions perpendicular to the aggregate surface. It constitutes thus a zone of weakness making easier the cracks propagation.
2. *Calcareous concretes*: the paste is compact and essentially constituted of portlandite (CaOH$_2$) and CSH gel . According to Evrard and Chloup-Bondant [1], the gypsium is incorporated in CSH gel, this explain the quasi-absence of ettringite.
Two types of contacts can be distingued: the first type is very compact with rough aggregates; it is possible to observe a difference between the transition zone and the cement paste. The presence of this layer can produce a modification of the microcracks type that become intragranular. In the second type, with smooth aggregates, the transition zone is less compact but continuous with the cement paste. Dissolution figures of the calcareous grains are observed. (photo 2).

4.1.2 - Samples after freezing and thawing cycles

The SEM observations allow to put in obviousness hydrates modifications [4]. The appearance of secondary ettringite (hexagonal prisms) is observed after freezing and thawing cycles (photos 3 and 4). The ettringite formation necessitates the presence of SO_4^{2-} ions, provide at the beginning of the hydration by gypsum :

$$6Ca^{2+} + 2Al(OH)^{4-} + 4OH^- + 3SO_4^{2-} + 26H_2O \rightarrow Ca_6[Al(OH)_6]_2(SO_4)_3.26H_2O$$

The freezing and thawing cycles entail the water circulation, provoking the apparition of many ions in the solution. Portlandite is dissolved and thus liberates Ca^{2+} ions. SO_4^{2-} ions are liberated from CSH gel in the case of calcareous concrete and by dissolution of primary ettringite in the case of siliceous concrete. The circulation of water allows the combination of the calcium, sulphate and silicate ions, causing the possibility of secondary ettringite formation.

The ettringite formation necessitates important quantities of water and the continuity of the liquid film governs the encounter possibility of elements ahead to react. The concrete compactness reduces movements of water, thus, if paste-aggregate contact is continuous and homogeneous, the possibility of ettringite formation decreases strongly.

Secondary ettringite formation entails an expansion of concrete provoked by the growth of needles. This expansion is facilitated with siliceous aggregates because of the existence of a free space between the paste and the aggregates, contrarily to the case of the calcareous aggregates. The expansion increases also if it exists already microcracks resulting from freezing and thawing cycles . The development of cracks after freezing and thawing cycles has been put in obviousness, however the observation of cracks on SEM must be deal with precaution because cracks can appear during the hydration or the preparation of samples. If the cracks contain new hydrates, that means that the cracks have appeared during the hydration or the freezing and thawing cycles (photo 5).

Ettringite crystallizes preferentially in free spaces [5], the presence of macropores can have a beneficial effect because these macropores can serve as chambers of expansion during the freezing periods (photo 6). After freezing and thawing cycles, the contact between cement paste and aggregates is difficult to observe because the aggregates are covered by new hydrates due to water movement.

The paste of Attigneville concrete is dense and the contacts are continuous: the circulation of water is therefore hindered, that explain the very few secondary ettringite development after freezing and thawing cycles.

Photo 1 : Siliceous transition zone
<-------------------> 18 µm

Photo 2 : Calcareous transition zone
<-------------------> 8 µm

Photo 3 : Ettringite in siliceous concrete
<------------------> 18 μm

Photo 4 : Ettringite in Gudmont concrete
<------------------> 8 μm

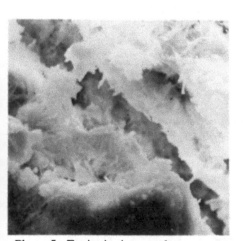

Photo 5 : Ettringite in a crack
Concrete of Jaumont
<------------------> 7,5 μm

Photo 6 : Ettringite in a bulk
Concrete of Gudmont
<------------------> 50μm

4.2 Physical and mechanical tests

The results of ultrasonic waves velocities, gas permeability and compressive strength, at total saturation, after 5, 25 and 300 cycles, are represented on Fig. 1, 2, 3. The majority of degradations of concretes appears before 25 cycles, indeed, few difference are observed between 25 and 300 cycles.

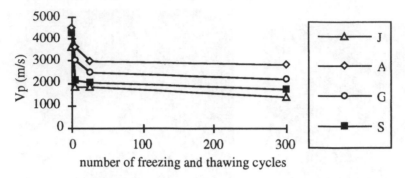

Fig. 1. Variation of ultrasonic waves velocities versus number of freezing and thawing cycles for total saturation

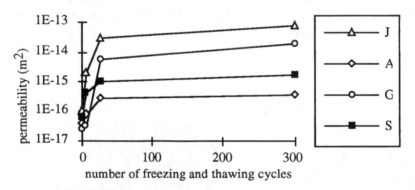

Fig. 2. Variation of permeability versus number of freezing and thawing cycles for total saturation

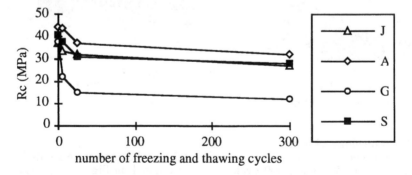

Fig. 3. Variation of compressive strength versus number of freezing and thawing cycles for total saturation

For siliceous concrete, the ultrasonic waves velocities decrease in majority before 5 cycles (Fig.1), also cracks appear rapidly, and the increase of gas permeability (Fig.2) indicates that cracks open mainly up to 25 cycles. The diminution of the compressive strength (Fig.3) shows a decrease corresponding to development and opening of cracks put in obviousness by ultrasonic waves velocities and gas permeability. The paste-aggregate contact is discontinuous , there is therefore initiation of microcracks at this level, which in fact prevails on the detachment of the aggregates.

The SEM observations of the Attigneville concrete show that very few secondary ettringite appears after the freezing and thawing cycles. The crystalline structure is hardly disturbed. The weak evolution of the all characteristics confirms these observations. Few cracks are created and they do not open, so that the compressive strength remains important. The transition zone is continuous and it is nearly possible to consider that in this concrete the transition zone and the cement paste can be considered as an alone porous network.

For Gudmont concrete, the diminution of the ultrasonic waves velocities (Fig.1) and the increase of the gas permeability (Fig.2) show that cracks initiate as early as the first cycles. The values of compressive strength (Fig.3) decrease fastly, showing that cracks open up to 25 cycles. The Gudmont aggregates are heterogeneous so that it exists strong contacts between cement paste and the oolithic aggregates because of their rough surface and the contacts between the sublithographic aggregates with smooth surface and the cement paste are less compact but more continuous. When the transition zone is very compact the water can not easily circulate during freezing and thawing cycles, this induces important excessive pressure and therefore the appearance of microcracks in the cement paste. Microcracks are going then to facilitate the circulation of water, therefore the appearance of secondary ettringite and important disorders.

For the Jaumont concrete the values of ultrasonic waves velocities (Fig.1) indicate the initiation and development of cracks and the values of gas permeability (Fig.2) show the opening of these cracks but the little decrease of the compressive strength values (Fig.3) shows that the cracks do not disturb very much the resistance of this concrete compared to the Gudmont concrete. The contacts between the cement paste and the aggregates are very strong, then the transition zone is more compact than the cement paste. Aggregates and their transition zone form a complex, like a biggest aggregate more resistant than the Jaumont aggregate. For this concrete, the cracks occur only in the cement paste, so that this concrete is less damaged than concrete of Gudmont.

5 Conclusions

This study shows that the damage of concretes during freezing and thawing cycles happen from the first cycles, so that the frost resistance test can be limited to 25 cycles. The SEM analysis has allowed to put in obviousness the appearance of secondary ettringite from the first cycles in damaged concretes. The creation of cracks improves the circulation of fluids, which facilitates the development of ettringite. Moreover, this ettringite entails an expansion of the concrete resulting from the crystalline thrust.

Complementary informations are obtained with ultrasonic waves velocities and permeability measurements. The compressive strengths give a confirmation about the damages. The tests are carried out with totally saturated samples because the mechanisms of degradation are more clearly identified. In relation with the type of aggregates and their states of surface three modes of damage are observed. No reactive siliceous aggregates lead to a lost of coherence between the cement paste and the aggregates, so cracks appear around the gravel. The calcareous aggregates with smooth surface react with the cement. The transition zone and the cement paste form a compact regular network. The crack formation is limited and the frost durability is

high. The calcareous aggregates with rough surface induce a very compact transition zone. The aggregates and its transition zone form a bigger aggregate more resistant than the initial aggregate. The cement paste represents a weak zone where the cracks growth.
The frost resistance results from several factors : presence of entrained air already mentionned by numerous authors, high cement paste-aggregate continuity depending on the roughness of the reactive calcareous aggregates. At least, the frost resistance of the calcareous concretes is not directly linked to mechanical performances of the aggregates but depends on the characteristics of the transition zone.

6 References

1. Evrard, O. and Chloup-Bondant, M. (1994) Réactivité chimique des calcaires en milieu basique : application aux ciments et bétons. *Annales de l'I.T.B.T.P.*, 529, pp. 83-87.

2. Chaouch, M. (1996) Etude des propriétés physiques et mécaniques de mortiers et bétons calcaires. Thèse INPL, Nancy, France.

3. Baron, J. and Ollivier, J.P. (1996) Les bétons. Bases et données pour leur formulation. *Ecole française du béton - Association technique de l'industrie des liants hydrauliques,* Eyrolles.

4. Pigeon, M. (1989) La durabilité au gel du béton. *Materials and structures*, 22, pp. 3-14.

5. Taylor, H.F.W. (1996) Ettringite in Cement Paste and Concrete. Séminaire RILEM, Béton "du matériau à la structure", 11 et 12 septembre 1996, Arles.

12 GRANITIC ROCKS ATTACK BY ALKALI–SILICA REACTION AND GELS FORMATION IN THE CEMENT PASTE INTERFACE

E. MENÉNDEZ
Instituto "Eduardo Torroja" de Ciencias de la Construcción, CISC, Madrid, Spain

Abstract
Alkali-silica reaction is not a common problem with regard to the Spanish aggregates, but in several conditions this kind of reaction is observed, with especial curing treatments. We have observed a concrete reaction in a pavement due to alkali-silica reaction. A general flaking was observed one year after the construction and some cones of material were flaking up. Granitic rocks presented a considerable reaction and a large quantity of gel formation was observed. A study of the reaction products was made by different techniques.
Keywords: Alkali-silica reaction, BSE microscopy, Concrete, Granite, SEM microscopy.

1 Introduction

A degradation process was observed in a concrete pavement one year after the construction. This pavement was built with granitic aggregates and ordinary Portland cement, this cement has a 0,8% of Na_2O equivalent. The Relative Humidity inside the construction was about 80%.

After the concrete slab was constructed, an isolating painting was applied on the surface of the concrete when this concrete was alredy wet.

The concrete showed a flaking process. The flack's sizes were about 5 mm, and were like some cones. In the vertexes of these cones there appeared some dark brown aggregates with a gelatinous aspect. Also, in some parts of the pavement a gel secretion could be observed.

The Interfacial Transition Zone in Cementitious Composites, edited by A. Katz, A. Bentur, M. Alexander and G. Arliguie. Published in 1998 by E & FN Spon, 11 New Fetter Lane, London EC4P 4EE, UK, ISBN: 0 419 24310 0

2 Experimental

Several cores of the concrete were studied by different techniques. The studied samples and the equipment used were the following:
* X-ray diffraction of concrete and of the attaced aggregates, with an equipment Phillips, Mod. PW 1710, with a copper tube with 40 kV and 50 mA.
* Infrared spectroscopy of the attacked aggregates and of the gel with an equipment Perkin Elmer, Mod. 783.
* Stereo microscopy of the fracture surfaces and of the cut surfaces, with a NIKON SMZ-27 equipment with an NIKON Microflex AFX-DX photographic system.
* Optical microscopy of the concrete and attacked aggregates using a microscopy ZEISS, Ultraphot II.
* Scanning electron microscopy and backscattering electron microscopy of the concrete with JEOL 5400 microscopy and with a microanalysis LINK of OXFORD, Mod. ISIS.

3 Results And Discussion

Different attacked granitic aggregates were observed by **stereo microscopy**, these aggregates had 20 millimeters size approximately and sized between 8 and 12 millimeters depth with respect to the pavement surface. Also, some gels inside the pores could be observed near of the attacked aggregates and there were some cracks going from the attacked aggregates to the surface of the pavement. These cracks had some gel inside.

In the **Figure 1** a granitic attacked aggregate with gel in the paste interface and two cracks going to the surface can be observe; also there is gel inside the cracks and in the pore near of the attacked aggregate. The cracks are development from the attacked aggregates inducing formation of cones and later the flaking process.

Attacked aggregates deeper than 15 millimeters approximately from the surface of the pavement could not be observed. However, similar' aggregates were found in other parts of the concrete.

The concrete was characterized by **X-ray diffraction** founded granitic compounds as the main kind of aggregates. The crystalline compounds of these granitic rocks are these: quartz, illite, montmorillonite, moscovite, albite and microcline. In **Figure 2**, we can observe the X-ray spectrum of the concrete.

Different thin sections of attacked aggregates and concrete were analyzed by **polarized optical microscopy** and we could observe that most of the altered aggregates were amorphous silicates with a little quantity of crystalline portions only. In the concrete thin section observation we could saw gels around an attacked aggregate and inside the cracks near to this aggregate. Also, some gels appeared inside of a pore near of the cracks and the attacked aggregate.

Different altered aggregate's portions were analyzed by **infrared spectroscopy** and mainly bands corresponding with the silicates were founded. In the **Figure 3** there is an altered aggregate infrared spectrum.

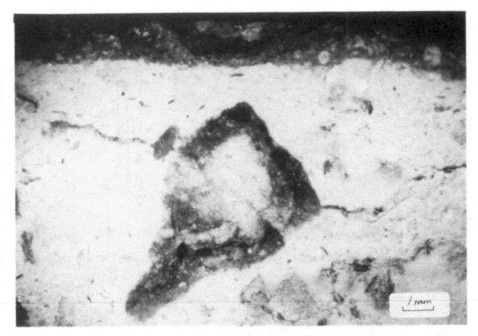

Figure 1. Optical microscopy of attacked aggregates and cracks going to the surface.

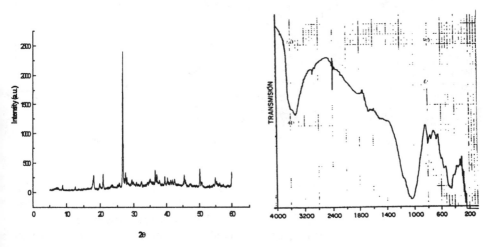

Figure 2. X-ray spectrums of the concrete.   Figure 3. An infrared spectrum of an altered aggregate.

3.1 SEM-EDX analyses

Alkali-silica reactions products could be identify on the fracture surface by scanning electron microscopy. These products were mainly massive gel rich in silicon, potassium and sodium **(Figure 4)** and needle like crystals rich in sodium **(Figure 5)**. These products appeared on the altered aggregates and, also massive gel was found on the paste in the fracture surface.

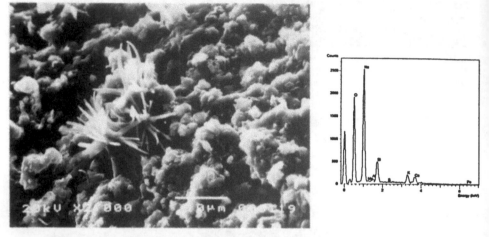

Figure 4. A SEM image of attacked aggregate with needle like crystals.

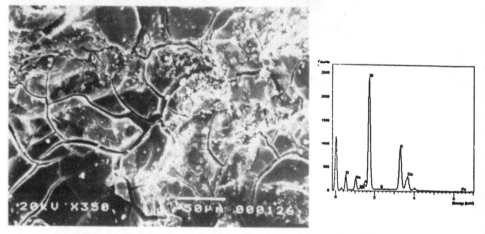

Figure 5. Massive gel on the fracture surface. SEM image.

3.2 BSE-EDX analyses

An attacked concrete area was observed, with an altered granitic aggregate and with a lot of gel formed in the interface. Also, some gel inside the cracks was formed near the attacked aggregate. And, some gel in a pore near the cracks was observed also.

In the interfacial zone between the paste and the altered aggregate, a lot of gel formation could be observed. This gel is rich in silicon and potassium and also, in sodium. Also, a crack formation with gel inside was saw from the aggregate **(Figure 6)**. In the **Figure 7** there are a series of X-ray dot map of the same area that has in the Figure 6. The massive gel in the paste-aggregate interface and in the crack is rich in silicon, in potassium and also in sodium, although some calcium signals can be observing mixed with the massive gel in the crack area. No ettringite evidences could be found in the interfacial zone of the altered aggregate.

The altered aggregate was compound fundamentally by siliceous material, but a rest of other component was found in the altered aggregate, this is a particle rich in potassium, silicon and aluminum. In the **Figure 8** there is an image of this area. A series of X-ray dot map of the previous Figure is in the **Figure 9**. The different areas of the altered aggregate can be seen . Also, the gel composition besides the aggregate zone has a high quantity of potassium and silicon, and less quantity of sodium. Nevertheless, the gel composition besides the cementicious paste has some points of calcium mixed with the silicon and potassium points.

Some X-ray point analyses had been taken from the gel in the paste-aggregate interface in order to known the possible compounds differences between the near paste and the near aggregate areas. In the **Figure 10** is point selection of the interfacial zone. The relative quantity of sodium is higher near the aggregate than near the paste, although the relative quantity of potassium and calcium is higher near the paste than near the aggregate. In the intermediate zone of gel is enrichment in silicon **(Table 1)**.

A different area of gel, inside the cracks and the pore, was study and some X-ray point analyses had been taken. In this area a crack in the paste with gel inside and a pore can be seen, also with gel inside **(Figure 11)**. By the point analyses we can found that the gel composition inside the crack is rich in silicon, and less in potassium and sodium, while the gel inside the pore has more quantity of calcium, may be due to the calcium aport of the portlandite of the pore. The paste composition near the pore is rich in calcium and not so much rich in silicon. Although the paste composition near the crack is more rich in silicon than near the pore. In general, the paste near gel' areas has a high quantity of potassium and sodium **(Table 1)**.

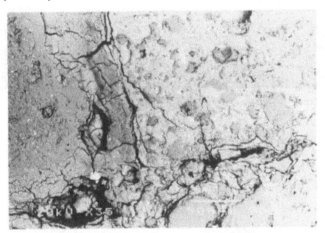

Figure 6. Gel formed in the paste-aggregate interface and inside the crack. BSE image.

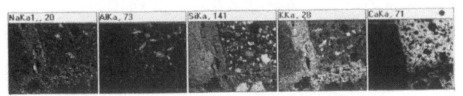

Figure 7. X-ray dot maps of the area shown in the Figure 6.

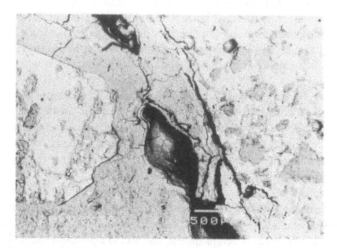

Figure 8. Attacked aggregate with gel formation. BSE image.

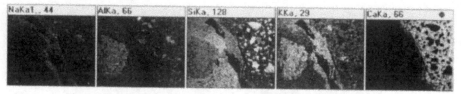

Figure 9. X-ray dot maps of the area shown in the Figure 8.

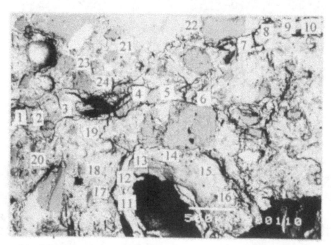

Figure 10. Massive gel in the paste-aggregate interface. BSE image.

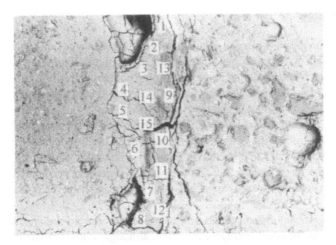

Figure 11. Massive gel inside of cracks and pores. BSE image.

Table 1. X-ray point analyses in percentage taken in figures 10 and 11

Figure 10				Figure 11						
Point	Na$_2$O	SiO$_2$	K$_2$O	CaO	Point	Na$_2$O	Al$_2$O$_3$	SiO$_2$	SO$_3$	K$_2$O	CaO
1	11.6	67.1	18.7	2.6	1	5.3	6.9	72.3	0	13	2.5
2	8.7	71.1	16.3	3.9	2	8.9	0.1	72.4	0	15.1	3.4
3	9.2	70.7	16	4.1	3	8.2	0	72.4	0.2	15.3	3.9
4	13	67.7	15.1	4.2	4	8	0	72.8	0.1	15.1	4
5	7.8	72	15.7	4.5	5	9.7	0	71.9	0	14	4.4
6	7.7	70	17.5	4.8	6	4.8	2.8	73.7	0.5	13.4	4.8
7	7.2	69.7	18.3	4.8	7	6.4	0	75.4	0	13.8	4.4
8	5.7	75.3	14.9	4.1	8	3.9	9	72.7	0.2	10.9	3.3
9	7.1	65.7	21.1	6.1	9	6.4	2.6	72.7	1.2	11.7	5.4
10	4.5	72.9	18.4	4.2	10	3.7	3	72.6	0.5	15.1	5.1
11	6.8	71.3	16.7	5.2	11	9.3	0	70.9	0.2	13.8	5.8
12	1.8	68.3	23.4	6.5	12	6	0	73.8	0.1	15.7	4.4
13	8.2	74	14	3.8	13	6.9	0	70.8	0.3	15.6	6.4
14	4.2	78.7	13.5	3.6	14	9	0	69.9	0.2	14.7	6.1
15	3.8	80.2	12.7	3.3	15	8.4	0	66	0	15.9	9.7
					16	9.2	1.7	68.1	0.2	15.6	5.2
					17	4.4	0.2	14.6	2.9	3.2	74.7
					18	3.5	1.6	16.2	5.5	3.1	70.1
					19	2.5	8.9	23.1	8.1	3.5	53.9
					20	1.2	7	15.9	2.5	9.6	63.8
					21	5.9	4.3	23.1	2.2	4.9	59.6
					22	4.8	3.7	23.1	4	3.7	60.7
					23	4.8	3.4	25	4.8	3.7	58.3
					24	3.2	2.8	26.8	2.1	9	56.1

4 Conclusions

- The preferential attack to the aggregates between eight and twelve millimeters like indicate that a deferential concentration of liquid phase could be produced due to the application of isolating painting with the concrete already wet.
- The gels progress toward the surface from the altered aggregates, this fact inducing the cracks' formation and the flaking process.
- The rest of the altered aggregates are formed mainly by amorphous silicates, this fact can be indicating a degradation process of the aggregates phases that contain potassium and sodium and consequently a liberation of alkalis to the medium.
- On the altered aggregate surfaces massive gel and needle like crystals could be observed. Also, massive gel appeared on the paste on the fracture surface.

- A significant difference has not been found in the gel composition in the different areas of the paste-aggregate interface, although a little enrichment in calcium could be observed near of the paste. The gel inside the cracks and inside the pores the quantity of alkalis is relativity less than in the paste-aggregate interface.

5 References

1. Imrani, B. and bernard, J.: "Use of gel composition as a criterion for diagnosis of alkali-aggregate reactivity in contrte containg siliceous limestone aggregate". Materials and Structure. Vol. 20, 1987
2. Menéndez, E.: "Microstructural study of alkali-silica reaction products in concrete cured at elevated temperatures". Materiales de construcción 43 (1993), 21-24
3. Thaulow, U., Jakobser, U.H. and Clark, B.: *Composition of alkali-silica gel and ettringite in concrete railroad ties: SEM-EDX and X-ray diffration analyses.* Cement and concrete research, Vol. 26, N°2,pp 309-318 (1996).
4. Menéndez, E.: "Modification of the alkali-silica reaction products microstructure by accelerated test in alkaline medium of different concretes". Sixth Euroseminar on Microscopy Applied to Building Materials. Reykjavik (Iceland). 25-27 June, 1997.

13 SULFATE ATTACK, INTERFACES AND CONCRETE DETERIORATION

J.P. SKALNY
Consultant, Holmes Beach, FL, USA
S. DIAMOND
Purdue University, Lafayette, IN, USA
R.J. LEE
RJ Lee Group Inc., Monroeville, PA, USA

Abstract
Sulfate attack is a generic name for a number of chemical and physical mechanisms of concrete deterioration resulting from uptake of external sulfate or sulfate-bearing solutions. It may lead to complete restructuring of the cement paste microstructure and chemistry at the paste-aggregate interfaces and elsewhere in the concrete. Illustrations of the modes of microstructural deterioration resulting from sulfate attack are presented, based on recent experience in the USA.

Keywords: concrete durability, deterioration, interfaces, microstructure, sulfate attack

1 Introduction

Sulfate attack is the term used to describe the complex mechanisms by which Portland cement-based concrete deteriorates by action of sulfate introduced into the interior of the concrete from internal or external sources. Cases of sulfate attack are well described in the classical scientific literature and were reported world over [e.g., 1,2]. In recent years, a number of new cases of external sulfate attack-related deterioration have been reported, leading to renewed interest in the details of the mechanisms involved [3,4].

In this paper we do not consider the effects of what in recent years has come to be known as "internal sulfate attack" or "delayed ettringite formation (DEF)" [e.g., 5,6], but confine our attention to effects produced from entry of external sulfate. As will be seen, the effects of external sulfate are themselves extremely complex, and in many respects are different from those commonly associated with DEF. Some new aspects of sulfate attack mechanisms will be highlighted, with emphasis on the influence of the source and form of sulfate, the concrete quality, and the role of the internal concrete interfaces.

The Interfacial Transition Zone in Cementitious Composites, edited by A. Katz, A. Bentur, M. Alexander and G. Arliguie. Published in 1998 by E & FN Spon, 11 New Fetter Lane, London EC4P 4EE, UK, ISBN: 0 419 24310 0

2 Processes Involved in Sulfate Attack

The processes leading to sulfate attack are caused by physical and chemical interaction of sulfate ions and accompanying cations with the components of the hardened cement paste matrix [1,2]; the physical and chemical phenomena occur simultaneously. However, to fully characterize a specific case of sulfate attack, one has to take into account the overall quality of the concrete (e.g., its water-to-cement ratio, degree of maturity, porosity and permeability) and its environmental exposure, especially the availability of external sulfates.

External sulfate attack may be induced by ingress of sulfate into hardened concrete from soil or ground water in contact with the concrete. Depending on the composition and ionic concentration of the external sulfate-bearing solution, various possible responses may occur, leading to a variety of physical-chemical scenarios. Note that some of these processes may be observed in a particular concrete in combination, this depending on the concrete quality, exposure conditions (e.g., the aggressive environment), and time. These individual or combined chemical processes of deterioration may or may not lead to significant expansion of the concrete matrix.

Chemical phenomena that may occur during external sulfate attack include:

- Release of aluminate ions from unydrated ferrite solid solutions (Fss, C_4AF) or other sources.
- Formation of ettringite in amounts beyond that previously formed from the sulfate present in the original cement.
- Removal of hydroxyl ions from the pore solution, leading to reduction of the pH.
- This pH reduction may be followed by dissolution and partial or complete removal of calcium hydroxide from the paste, leading to further decrease in alkalinity.
- Local decalcification of the calcium silicate hydrate (C-S-H), resulting in an altered microstructure and loss of binding capacity.
- As a consequence, some of the C-S-H is converted to a hydrated silica gel material, or sometimes to hydrous magnesium silicate.
- Dissolution of much of the residual unhydrated alite and belite in larger cement grains, and local replacement of the dissolved material by various secondary substances, including ettringite.
- Deposition of ettringite in open spaces such as pores and cracks, and around aggregate interfaces.
- Formation of gypsum.
- Formation of magnesium hydroxide (brucite).
- Formation of local deposits of calcium carbonate, i.e. carbonation, in areas of concrete in contact with the ground water.

Under otherwise constant conditions, the reaction products of chemical sulfate attack vary with concentration of the sulfate in the sulfate source, and depend on the accompanying cations (e.g., the proportions of Ca^{2+}, Mg^{2+} and Na^+), on the initial chemistry of the concrete pore solution, and on variations in exposure over time (e.g., alternations of wetting and drying, temperature variations), etc. The interaction of sulfate attack with carbonation and chlorides may play a significant role.

In addition to the various chemical effects, concrete may undergo damage by a form of *physical* sulfate attack, involving the effects induced by repeated alternate re-crystallization of sodium sulfate 10-hydrate (mirabelite) and sodium sulfate (thenardite), caused by repeated reversal of temperature and humidity. Such repeated re-crystallization in the pores of the affected concrete may lead to volume expansion and, possibly, to spalling and scaling [e.g.,4].

These internal chemical and physical responses result in modifications of the physical properties of concrete. The most important change is a progressive increase in porosity and permeability, causing acceleration of deterioration in the later stages of the process. The deterioration is often marked by expansion of the paste and the development of local cracks, and by loss of cohesion between paste and the aggregate. Spalling of exposed surfaces of the concrete is common, and is associated with sodium sulfate deposition and repeated re-crystallization of thenardite and mirabelite. In the later stages of deterioration visible cracking occurs in many areas of the affected concrete.

The overall strength of concrete is obviously affected, and in the final stages of the processes of sulfate attack may result in complete loss of structural integrity. However, there are complications in relating measured strength to the deterioration process. The deterioration processes do not take place uniformly through the concrete section, but the effects generally occur first near the point of entry of the sulfate-bearing solutions, and move upwards/inwards in layers with time. Concrete strength tests require large specimens that are at least several times thicker than the coarsest aggregate piece, and the height-diameter ratio must be within established limits. Thus the trend of concrete strength does not necessarily reflect the progressive deterioration taking place at the microstructural level. As clearly stated by Mehta [7], concrete strength per se is not an adequate measure of durability or lack of it.

Actually, the effects of various processes on strength are much more adequately reflected in tests of smaller mortar or paste specimens, such as have been used in research studies and reported by various authors. In general, exposure to sulfate solutions produces a slight increase in compressive strength at first; subsequently, a reversal occurs and progressive strength loss is measured as the degradation processes progress through the thickness of the specimens.

3 The Importance of Interfaces

Interfaces in Portland cement concrete are complex and, as is well known, their "quality" depends on the cement, aggregates, and mineral and chemical admixtures used, on processing conditions (controlling the *maturity* of the concrete), and on the environment to which the concrete is exposed. It appears that the controlling feature in the resistance of concrete to various processes in sulfate attack is the relative ease with which the attacking solution components can be absorbed into and transported through the pores of the concrete. Other factors, including the solid mixture proportions, the type of cement, and even the concentration of the external sulfate source have very much less influence than the permeability of the concrete or, more specifically, the pore structure and connectivity of the larger (capillary) pores.

The ease with which the attacking solution components can be absorbed and transported through the pores of the concrete is well known to be a function of the water-to-cement ratio used. For water-to-cement ratios much in excess of 0.45 the retention of a permeable, interconnected pore structure is inevitable, regardless of the extend of hydration. This well-known fact is reflected, for example, in the requirement of the U.S. Uniform Building Code: concretes placed in high-sulfate environments are required to be placed at water-to-cement ratios no higher than 0.45 [8].

According to current ideas, much of the internal transport of dissolved components through concrete may take place through the more porous and interconnected interfacial transition zones (ITZs) surrounding aggregates and sand grains. This may be less true for high water-to-cement ratio concrete than for less permeable concrete mixed at lower water contents. To the degree that it is true, in high water-to-cement ratio concrete exposed to sulfate-bearing solutions the interfacial zones may play some special role in the responses produced. Furthermore, the role assigned to ITZs is expected to be augmented as progressive damage occurs. Progressive dissolution of calcium hydroxide and decalcification of the C-S-H may be expected to occur initially along solution transport paths, i.e. initially around interfaces, at least, to the extent that these predominate. Furthermore, local expansion of the matrix away from aggregate contact, as occurs in some phases of sulfate attack, produces a significant augmentation of the importance of the interface as a preferred channel for transport of ions and solutions.

4 Illustrations

In most reported cases of external (and internal) sulfate attack the concrete involved has been placed at relatively high water-to-cement ratios and more often than not, inadequately cured or not cured at all.

To provide a hypothetical example, consider two concretes, both made with ASTM Type V cement and both exposed to wet sulfate-bearing soils containing more than 1500 ppm of dissolved sulfate. This constitutes "severe" sulfate exposure according to ACI Committee 201 and Uniform Building Code [8]. One of the concretes is batched at a water-to-cement ratio of 0.65, the other one at 0.45. Sulfate solutions will rapidly penetrate the former concrete and by the processes of advection will be drawn through it, react with the paste components to form new compounds, and deposit crystalline sulfate salts at or near its upper surface. While some sulfate solutions will penetrate the w/c 0.45 concrete, if it has been adequately cured, the pores will be mostly discontinuous and the bulk of the concrete will act as a barrier to further penetration. The former concrete will be subject to progressive sulfate attack through the mass of the concrete; the latter will not.

In the illustrations that follow, we document some of the changes in microstructure and morphology induced by the internal processes of sulfate attack in exposed, high water-to-cement ratio concretes. The micrographs shown are for illustration only and they do not necessarily represent any particular case of sulfate-induced damage. In each case the micrographs have been produced in backscatter-mode SEM. Figs. 1-3 are individual micrographs only. Figures 4-8 are sets in which a low-magnification micrograph is shown on the left, each such figure containing an area outline by a square box. The area within the box is shown at much higher magnification in the micrograph at the right. Within this, a local feature selected for energy-dispersive X-ray (EDX) spot analysis is shown by a very small square. The EDX spectrum produced at that spot is reproduced below the micrographs.

Figure 1 represents concrete made with w/c = 0.65 and Type V cement, exposed to severe levels of sulfates in ground water. Note the high porosity, an obvious conduit for the pore solution.

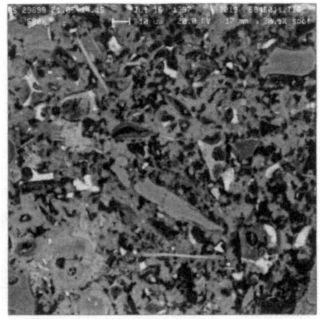

Figure 1

Figure 2 shows deposits of ettringite in the paste and at the interfaces between aggregate and the paste. Such deposits are typical for the calcium sulfate attack – meaning reaction of sulfates with the calcium and aluminate sources in concrete to form ettringite, possibly leading to expansion of the paste and severe cracking; similar phenomena are found as a result of internal sulfate attack, e.g. DEF.

Figures 3a and 3b illustrate the formation of veins of gypsum at interfaces and in bulk paste; his represents damage caused by high concentration of sulfates in the concrete pore solution.

Figures 4 to 6 show pseudomorphs of unhydrated clinker minerals in which the calcium ilicates, alite and belite, are completely decalcified (fig.4) or replaced by brucite (figs.5a and 5b) or hydrated magnesium silicate (fig.6). Presence of such pseudomorphs may indicate early xposure of the particular concrete to magnesium sulfate solutions even before the clinker calcium ilicates had a chance to hydrate.

Figure 7 represents an example of deposits of sodium sulfate in partially decalcified former ilicious cement particle. Note the relative ratios of the Si and Ca peaks in the EDS pattern, ndicating decalcification of C-S-H.

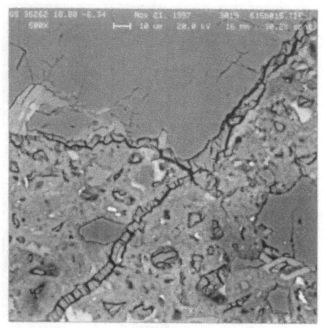

Figure 2

Figure 3a

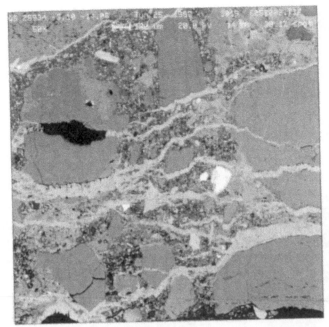

Figure 3 b

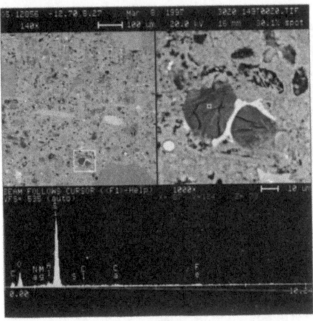

Figure 4

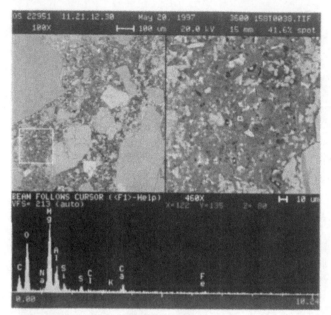

Figure 5a

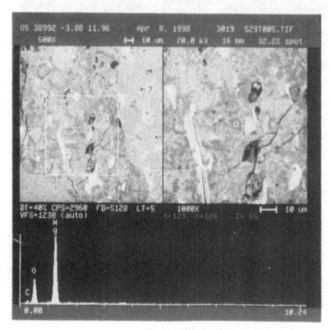

Figure 5b

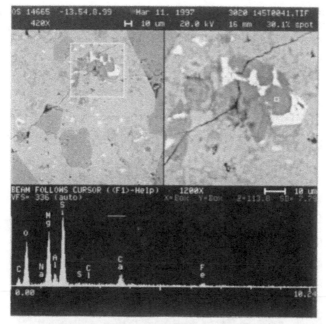

Figure 6

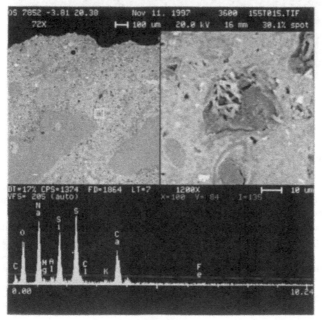

Figure 7

Finally, figure 8 shows presence of Friedel's salt (a calcium chloroaluminate hydrate) in a paste with severe deposits of ettringite. Such finding is characteristic of exposure of concrete to ground waters containing both sulfates and chlorides in appreciable concentrations.

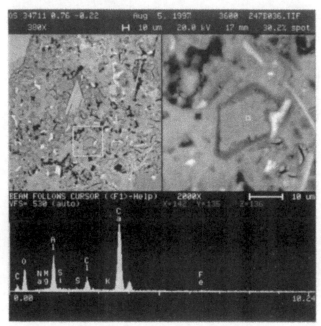

Figure 8

5 Concluding Remarks

All of the micrographs shown represent internal microstructural changes that occur in concrete undergoing sulfate attack. Some of them reflect changes along aggregate grains or within the generally-accepted (assumed) thickness of aggregate-induced ITZs; others do not. Certainly, to the extend that ITZs offer a preferred pathway for sulfate ions to penetrate to a particular region of the concrete, they may influence the progress of the sulfate attack. However, as illustrated, the various alterations that constitute the indications of sulfate attack processes are widely dispersed within the paste in a particular zone and not necessarily confined to the vicinity of the aggregates.

This generalization may stem from the overwhelming interconnectedness of the pore system in high water-to-cement ratio concretes in which these sulfate attack problems have been studied. Available research data show that at low water-to-cement ratios sulfate-related damage is confined to the outer few millimeters of the exposed specimens. Interfacial transition zones may be more important in cases where the general interconnectedness of the pores in cement paste is less complete than in these concretes.

The lesson to be applied for concretes to be used in hostile chemical environments is clear. Low water-to-cement ratios and proper curing practices are required to limit the degree to which external solutions can penetrate and follow the interconnected pore channels through the concrete. The use of ASTM Type V cement to limit the amount of ettringite formed is certainly helpful in some cases, but is no panacea in the absence of the design and production of effectively impermeable concrete. Even for a properly cured concrete the use of water-to-cement ratios as low as 0.4 should not be ruled out for applications where extremely high levels of sulfate ions may occur.

References

1   Taylor, H.F.W.(1997), *Cement Chemistry*, 2nd Edition, Thomas Telford; see also:(1996), Ettringite in Cement Paste and Concrete, presentation at the conference "Beton: du materiau a la structure", Arles, France, in press; see also: (1994) Sulfate Reactions in Concrete – Microstructural and Chemical Aspects, in *Cement Technology* (E.M.Gartner and H.Uchikawa, Eds.), The American Ceramic Society, pp.61-78
2   DePuy, G.W. (1994), Chemical Resistance of Concrete, *in Tests and Properties of Concrete* (P.Klieger and J.F.Lamond, Eds.), STP 169C, ASTM, Philadelphia, pp.263-281
3   Mehta, P.K. (1992), Sulfate Attack on Concrete – A Critical Review, *in Materials Science of Concrete*, Vol. III (J.Skalny,Ed.), The American Ceramic Society, pp.105-130
4   Haynes, H., O'Niell R, and Mehta, P.K., Concrete Deterioration from Physical Attack by Salts (1996), *Concrete International*, Vol.18, No.1, pp.671-677
5   Lawrence, C.D. (1995), Delayed Ettringite Formation: An Issue?, *in Materials Science of Concrete*, Vol. IV (J.Skalny and S.Mindess, Eds.), The American Ceramic Society, pp.113-154
6   Taylor, H.F.W. (1994), Delayed Ettringite Formation, *Advances in Cement and Concrete*, ASCE, New York, pp.122-131
7   Mehta P.K. (1997), Durability – Critical Issues for the Future, *Concrete International*, July 1997, Vol.19, No.7, pp.27-33
8   Uniform Building Code, 1988 Edition, Table 26-A-6, p.516; see also Guide to Durable Concrete, 199 , section 2.2, ACI 201

14 INTERFACIAL EFFECTS IN GLASS FIBER REINFORCED CEMENTITIOUS MATERIALS

K. KOVLER and A. BENTUR
Technion – Israel Institute of Technology, Haifa, Israel
I. ODLER
Technical University of Clausthal, Clausthal, Germany

Abstract
The performance of both E-glass and AR-glass fiber reinforced composites with four low alkali/low lime cementitious matrices made of glass, calcium aluminate phosphate, magnesia phosphate and ettringite cements, was studied. After testing, SEM observations were carried out to resolve the mechanisms involved in the aging process, in particular those of microstructural origin (i.e. the densening of the matrix at the interfacial zone), and those of chemical origin (i.e. those which caused defects to form on the fiber surface).
Keywords: Glass, cement, fiber-reinforced composites, aging, durability, SEM.

1 Introduction

In recent years considerable efforts have been directed at the development of new glass fiber reinforced cementitious composites intended for replacement of asbestos-cement, as well as to generate new materials of improved properties, in particular toughness. However, due to the high surface area (most of the modern glass fibers are made in the form of thin filaments, 10 to 20 μm in diameter) the fibers are sensitive to interfacial effects. Such effects can be chemical or physical in nature, and beneficial or deleterious from the durability viewpoint.

The beneficial effect is the enhanced bond. The deleterious effect can be the consequence of the development of a bond, which is too strong, leading to embrittlement, or chemical attack, especially when using Portland cement matrix, which is highly alkaline. The glass is particularly sensitive to aging effects since it tends to corrode in high alkalinity matrix, and even if its composition is made

The Interfacial Transition Zone in Cementitious Composites, edited by A. Katz, A. Bentur,
M. Alexander and G. Arliguie. Published in 1998 by E & FN Spon, 11 New Fetter Lane,
London EC4P 4EE, UK, ISBN: 0 419 24310 0

immune to alkaline attack, it may fail readily in a brittle mode when microstructural densening occurs due to deposition of $Ca(OH)_2$ around the glass filaments.

The matrix densening around the thin glass filaments may cause embrittlement by changing the nature of the reinforcing strand from a flexible unit, in which one filament can slide over the other, and can thus accommodate local stress concentrations, into a rigid unit which enables crack propagation through the bundle [1].

In view of this mechanism, various means were developed to prevent interfacial densening and consequently improve long-term performance. These means included the development of special fiber formulations to be compatible with the cement matrix, treatments of the fiber surface to prevent nucleation of $Ca(OH)_2$ [2], impregnation of the spaces between glass filaments in strands by polymer [3] and silica fume particles [4] to prevent local growth of $Ca(OH)_2$.

The approach based on treatments of the fibers can be successful, but it is frequently costly to develop and implement in production, unless a large market is guaranteed in advance. A different approach, to modify the overall composition of the matrix to prevent formation of $Ca(OH)_2$ and preserve a more porous matrix was suggested recently [5].

In the previous study of the authors [6] it was suggested to use specially developed low lime - low alkalinity cement matrices for such composites. Four low alkali/low lime cements were developed recently at the Technical University - Clausthal: glass (GC), calcium aluminate phosphate (CAPC), magnesia phosphate (MPC) and ettringite (EC) cements, which have the potential for providing a matrix for high performance composites. This approach may provide a more general solution to accommodate the fibers, since many of the problems cited previously are common to the different fibers incorporated in the cement matrix; they are associated with the high alkalinity of the Portland cement (PC) and its tendency to develop large, dense deposits of $Ca(OH)_2$ in the vicinity of the fibers. These two effects can simultaneously be taken care of by developing low alkali/low lime cements, which could be economically feasible, since they can be made of various industrial by-products.

In the present study the performance of both E-glass and AR-glass fiber reinforced composites with these cementitious matrices was studied. The results were compared with those of ordinary Portland cement (PC) composites.

2 Experimental

2.1 Materials
The first cement matrix studied was GC. It has been reported that some glasses in the system $CaO-Al_2O_3-SiO_2$, if ground to a high fineness, exhibit cementitious properties. The products of the hydration process are hydrogarnet $C_3AS_xH_{(6-2x)}$ and stratlingite (gehlenite hydrate) C_2ASH_6. Cements of this type may be considered candidates for GRC matrices due to both the relatively low pH value (11.0) of the pore solution and the absence of calcium hydroxide in the hydration products.

The second cement matrix chosen was CAPC. The binder was obtained by mixing high alumina cement with sodium polyphosphate $(NaPO_3)_n$. This cement was also included in our study as it does not liberate Ca(OH$_2$) (portlandite) in its hydration and exhibits a low pH value (11.4).

The cement matrix (MPC) was prepared on the basis of a blend of magnesium oxide with concentrated solutions of diammonium phosphate. Since the liquid phase in equilibrium with the reaction product has a relatively low pH value (11.2), and since calcium hydroxide is not among the products of reaction, this binder was the third one to be included in our study as a possible candidate for GRC composites.

The fourth non-Portland cement was an entirely new low pH, Ca(OH)$_2$ free binder. The binder is a blend of a CaO-Al$_2$O$_3$-SiO$_2$ glass (80%) and gypsum (20%). Its hydration product is ettringite. Usually, the ettringite phase forms partially in a topochemical and partially in a through-solution reaction. In such pastes, expansion occurs during the hydration. In the hydration of a paste made from a CaO-Al$_2$O$_3$-SiO$_2$ glass and gypsum, the ettringite formation took place exclusively in a through-solution process and was not accompanied by an expansion. The pH value of the liquid phase is 11.0.

2.2 Testing

The performance of the glass fibers in the cementitious matrix was evaluated by testing glass fiber-cement composite specimens in flexure. The specimens were beams, 110x20x10 mm in size, in which continuous strands of glass fibers were placed in the bottom part, 3.3 mm from the lower face. The specimens were prepared in special molds, where a 3-mm thick layer of paste was first cast. Over it, continuous strands were placed and impregnated with paste, and on top of them a second layer of cement paste was cast. The total thickness of the specimen was 10 mm, and the volume content of the glass fibers was about 1%. In each mold, six beams were prepared.

The fiber reinforced beams were demoulded after one day, and then cured in water at 20°C until 28 days. The properties at this age represented those of unaged composites.

A hot water test (immersion in 60°C water) constituted the accelerated aging. Accelerated aging commenced after 28 days of 20°C water curing and continued for periods of up to 56 days, with tests being carried out at 0, 14, 28, and 56 days of accelerated aging.

The performance of the composites was evaluated by a flexural test, performed by four point loading at a span of 90 mm, and at a 0.5 mm/min bridge cross-head movement. The load and mid-span deflection were continuously recorded, and load-deflection curves were obtained. From the curves, the following parameters were derived: first crack stress or limit of proportionality (LOP), representing the matrix strength, maximum flexural stress in the post-cracking zone, and the area under the curve up to a deflection of 1.6 mm.

The specific fracture energy was calculated as the area under the load-deflection curve, relative to the cross-section. The change in the specific fracture energy or in the post-cracking stress as a function of the period of accelerated aging served as a means for evaluating the durability performance.

After testing, SEM observations were carried out to resolve the mechanisms involved in the aging process, in particular those of microstructural origin (i.e. the densening of the matrix at the interfacial zone), and those of chemical origin (i.e.

those which caused defects to form on the fiber surface). The SEM observations were carried out on the fractured surface exposed in the flexural test, as well as the surfaces exposed by splitting of the specimen in the plane of the fibers.

3 Mechanical properties

3.1 E-glass composites

The four different low alkali/low lime matrices were developed and their potential for producing high durability glass fiber reinforced cementitious composites were studied in E-glass fiber reinforced composites. This glass is particularly sensitive to aging effects since it tends to corrode in high alkalinity matrix, and due to its brittle nature, it may break readily when microstructural densening occurs. It thus provides a good reference to study the efficiency of the special matrices in enhancing durability.

It was found that the PC had the highest strength matrix (LOP was 14.6 MPa before aging), and MPC and EC had the lowest ones (5.9 and 5.3 MPa, respectively). The strengths of GC and CAPC were intermediate (7.0 and 9.8 MPa, respectively). The strengths of the investigated cements had a tendency to increase by an average of 1-3 MPa during the period of aging, except the EC and PC, which showed no increase during water-saturated curing.

The materials studied may be classified according to their mechanical performance by means of either post-cracking flexural strength, or fracture energy: based on these criteria, the best durability performance was obtained with MPC and EC matrices. The typical load-displacement diagrams of the MPC composites are shown in Fig. 1. The worse performance was observed with CAPC and PC. The GC matrix was intermediate.

Only in two cases the mechanical properties of E-glass fiber reinforced cement were improved during aging: it was with MPC and EC. In the three other cases, the mechanical performance of the composite in the post-cracking zone was already weakened at the initial aging period. For aging longer than 14 days, the post-cracking flexural strength and fracture energy did not change dramatically.

3.2 AR-glass composites

The four cement matrices (GC, CAPC, MPC and EC) were evaluated also with AR-glass fibers. Here, the significance of the matrix, from the point of view of microstructural effect, is to be dominant, as the fibers are much more immune to chemical attack.

It was found that in the AR composites, the GC matrix provided the best aging performance, eliminating strength reduction and even providing enhancement with age. The MPC, which provided the best performance in the E-glass system, was second best here. Both systems performed much better than the AR-glass with PC and CAPC matrix. The ettringite cement did not reveal any significant decrease in the post-cracking and can be considered as well as the second best. For the comparison with the mechanical behavior of the E-glass composites made from the same matrix (MPC and EC), the load-displacement diagrams of the MPC and EC AR-glass composites are shown in Fig. 2.

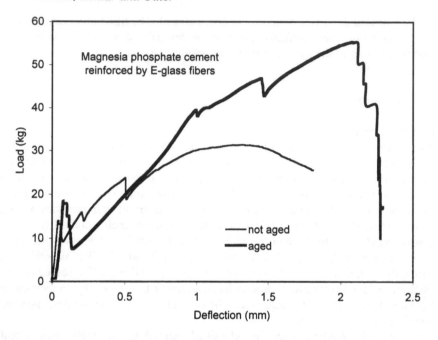

Fig. 1. Typical load-displacement diagrams in flexure of beams from magnesia phosphate cement, reinforced by E-glass fibers, obtained before and after 28 days accelerated aging.

4 Microstructure

4.1 E-glass composites

Consistent SEM observations of five matrices reinforced with E-glass fibers were carried out after 28 days of 20°C curing and after 28 days of accelerated aging at 60°C.

Prior to aging of GRC specimens with PC matrix, a little penetration of hydration products into the glass fiber strand was observed; the matrix in the vicinity of the strand and between the filaments was quite porous. After 28 days of accelerated aging, there was some densening at the interfaces, with more growth of hydration products observed between the filaments. There was at this age, in addition, marked signs of considerable etching of the fibber surface, including peeling of external layers in the fibers, indicative of a severe chemical attack.

After 28 days of accelerated aging of GRC specimens with GC matrix, the matrix around the glass filaments and between them was quite porous (Fig. 3), similar to the observations prior to aging. However, considerable signs of chemical attack could be seen, exhibited as relatively large pits on the glass surface, as well as roughening of the surface.

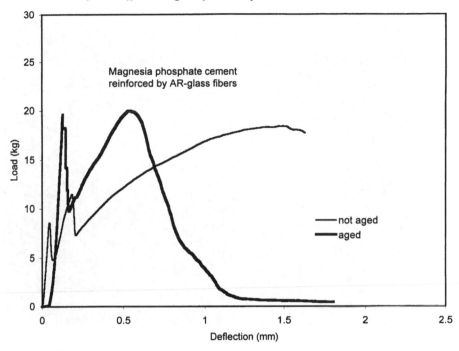

Fig. 2. Typical load-displacement diagrams in flexure of beams from magnesia phosphate cement, reinforced by AR-glass fibers, obtained before and after 28 days accelerated aging.

The superior aging behavior of MPC matrix reinforced by E-glass fibers was consistent with microstructural observations, showing no signs of chemical attack on the fibers as well as a matrix which, although dense, was readily peeled off the fibers. As in the case of the MPC matrix, the EC matrix did not show any sign of chemical attack on the fibers. The structure of the ettringite cement matrix remained friable, and the formation of ettringite needles did not lead to any damage of E-glass fibers. Considerable signs of chemical attack were observed in the systems with matrices of PC, GC, and CAPC. All these systems exhibited considerable loss in post-cracking strength and toughness. The systems with MPC and EC matrices, which did not lose strength or toughness, were characterized by the absence of any signs of chemical corrosion of the fibers. Thus, in the E-glass composites, the durability performance could be readily correlated with immunity to chemical attack, and the major secondary influence of the microstructure effects. In the systems, which exhibited loss in post-cracking strength and toughness, chemical attack was readily observed (GC, PC and CAPC). Of these three, the loss in mechanical performance was at the slowest rate in the GC system. This difference might be correlated with microstructural influence: the microstructure was very dense in the PC and CAPC, but much more porous in the GC composite.

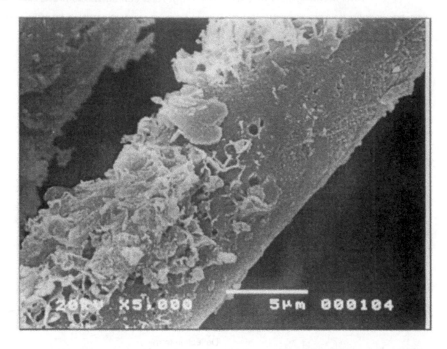

Fig. 3. Micrograph of E-glass composite with GC matrix after accelerated aging.

The superior behavior of the MPC and EC systems was correlated with their immunity to chemical attack. Both systems exhibited a very porous interfacial microstructure, which is expected to be deficient in terms of providing sufficient bonding to mobilize the strength of the fibers. This showed up in the low first crack stress of these composites. On aging, some additional densening was obtained with the MPC system, which may account for the increase in strength of this system with aging. Apparently, this increase was sufficiently high to improve the strength of the composite, but it was probably still too low to cause premature fiber breakage by too high a bond or by local flexural failure.

4.2 AR-glass composites

The hydration products of PC densely adhered to the filament surface and penetrated in-between the adjacent filaments. A very dense matrix was observed, but there were no signs of corrosion in the glass surface.

Micrographs of the GC composite also showed clean AR-glass fibers, without any indication for corrosion (Fig. 4). However, the matrix was very porous.

The CAPC composite was characterized by the absence of signs of chemical attack on the fibers. Some of the filaments seemed to be covered with a thin layer of hydration products, and the matrix seemed as rather dense.

The AR-glass fibers in the vicinity of MPC matrix had no signs of corrosion. This fact could be expected because the E-glass fibers in the same matrix also did not indicate a chemical attack. The matrix looked dense, but less dense than in PC and CAPC. As far as EC composite is concerned, there were no signs of chemical attack. The matrix looked very friable and porous.

Fig. 4. Micrograph of AR-glass composite with GC matrix after accelerated aging.

5 Conclusions

Considerable differences in the aging of E-glass and AR-glass composites with different cementitious matrices could be observed and accounted for on the basis of microstructural and chemical effects. These results indicate that by adjusting the composition of the matrix, there is a potential for developing highly durable fiber-cement composites.

6 References

1. Proctor, B. A. (1990) A review of the theory of GRC. *Cement & Concrete Composites*, 12, pp. 53-61.
2. Proctor, B. A., Oakley, D. R. and Litherland, K. L. (1982) Development in the assessment and performance of GRC over 10 years. *Composites*, 13, pp. 173-9.
3. Bijen, J. (1990) Improved mechanical properties of glass fiber reinforced cement by polymer modification. *Cement & Concrete Composites*, 12, pp. 95-101.
4. Bentur, A. (1989). Silica fume treatments as a means for improving the durability of glass fibre reinforced cement. *J. Mater. in Civil Engng*, ASCE, 1, pp. 167-183.
5. Kuroki, K. and Hayashi, M. (1993) Durability of GFRC bending and fiber/matrix boundary microstructure, in *Durability of Building Materials and Components 6*, (ed. S. Nagataki, T. Nireki, F. Tomosawa), E & FN Spon, London, pp. 129-38.
6. Kovler, K., Bentur, A. and Odler, I. (1995) Durability of E-glass fibre reinforced composites with different cement matrices, in *Concrete under Severe Conditions*, Proc. 1st Int. Conf., (ed. K. Sakai, N. Banthia, O.E. Gjorv), E & FN Spon, Sapporo.

PART FIVE

ITZ AROUND REINFORCING BARS AND ITS ROLE IN STEEL CORROSION

15 ACTION OF CHLORIDE IONS ON THE REACTIONS IN THE CORRODED STEEL–CEMENT PASTE INTERFACIAL TRANSITION ZONE

J.L. GALLIAS and R. CABRILLAC
Laboratoire Matériaux et Sciences des Constructions, Université de Cergy-Pontoise, France

Abstract
The formation of hexagonal and cubic calcium ferrite hydrates in the interfacial transition zone between corroded steel and cement paste seems to be a very important factor for the improvement of physical and mechanical characteristics of the bond between concrete and reinforcement. The results of the study show that the chloride ions in the cement paste accelerate significantly the crystallization of the hexagonal calcium ferrite hydrate. With the increase of the concentration of chloride ions in the ITZ a complete series of solid solutions between calcium ferrite hydrates and calcium chloroferrite hydrates is formed. These new crystalline products fix an important part of free chloride ions in the ITZ and could reduce consequently the risk of chloride induced corrosion for the embedded reinforcement.
Keywords: Interfacial transition zone, corroded reinforcement, corrosion layer, cement paste, microstructure, calcium ferrite hydrate, chloride ions, calcium chloroferrite hydrate, solid solution.

1 Introduction

In a previous study [1], the authors concluded that the corrosion layer on steel combines a physical action before setting and a chemical action during cement hydration, when embedded in cement paste. First, high open porosity of the corrosion layer absorbs mixing water of the fresh paste and releases air, creating air bubbles in the limit of the corrosion layer and the cement paste. Second, the steel corrosion products of the outer part of the corrosion layer (full of iron hydroxides and oxyhydroxides) react with the absorbed calcium hydroxide of the fresh cement paste and produce hexagonal calcium ferrite hydrate. With aging, cubic calcium ferrite hydrate is formed in part as a substitution of the hexagonal calcium ferrite hydrate.

The Interfacial Transition Zone in Cementitious Composites, edited by A. Katz, A. Bentur, M. Alexander and G. Arliguie. Published in 1998 by E & FN Spon, 11 New Fetter Lane, London EC4P 4EE, UK, ISBN: 0 419 24310 0

The present study concerns the action of chloride ions on the microstructural phenomena in the ITZ, taking in account that the chloride ions could react with the cement components and could depassivate the embedded steel.

2 Materials and methods

2.1 Experimental model
The simple experimental model of the previous study was used to determine the action of chloride ions on the ITZ around corroded reinforcement :

- Small cylindrical pieces of 15 mm height and 20 mm diameter, cut from a typical mild steel reinforcing bar and burnished up to 5 µm abrasive, were exposed without protection for fourteen months in outdoor weathering.
- At the end of the exposure period and after one day room curing, one of the circular surfaces of the steel pieces were embedded into a fresh paste of ordinary Portland cement (CPA 55R according to NF P 15 301 French standard) mixed with a water-cement ratio of 0.40.
- Three series of corroded steel - cement paste composite specimens were prepared with additions of 0%, 1% and 3% $CaCl_2.2H_2O$ of the mass of the cement to the mixing water (corresponding to 0.0%, 0.48% and 1.44% Cl⁻ content in the paste).
- The composite specimens were cured at a relative humidity of $95 \pm 5\%$ and at a temperature of $20 \pm 2°C$, protected from carbonation.
- The separation of the composite specimens for observation was carried out by shear stress acting tangentially on the steel surface. In these conditions, the fracture occurs normally into the ITZ.

In addition to the composite specimens, pure goethite - portlandite - calcium chloride dihydrate - water mixtures were prepared using varying weight ratios from 1/5/0/5 to 1/5/5/5.

2.2 Characterization of composition
X ray diffraction (XRD) was used for the observation of the ITZ of the composite specimens and of the mixtures. In the case of composite specimens, XRD was applied on successive surfaces obtained by dry polishing on 5 µm abrasive in a parallel to the fracture direction, up to the uncorroded steel surface and up to the center of the cement paste. The measure of the removed mass and of the thickness variations at any step of the abrasion process, enabled us to determine the thickness and the mean apparent density of the whole ITZ of the composite specimens. The observations were carried out from 3 days after casting till 500 days.

3 Results and discussion

3.1 Effect of chloride ions on the hydration of the cement paste
The figure 1 shows the XRD diagrams in the center of the cement paste with various amounts of calcium chloride.   The more important modification on crystalline compounds is the gradual transformation of the calcium monosulphoaluminate hydrate $(3CaO.Al_2O_3.CaSO_4.12H_2O)$, with a main XRD peak at 9.0 Å, into the calcium

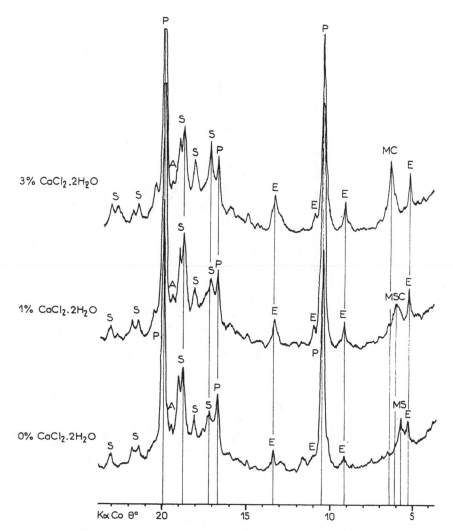

Fig. 1. X-ray diffraction diagrams of 72 days old cement pastes with various amounts of calcium chloride hydrate. S, calcium silicates (C_2S and C_3S); A, calcium aluminate (C_3A); P, portlandite E, ettringite; MS, calcium monosulphoaluminate hydrate; MSC, calcium sulphochloroaluminate hydrate; MC, calcium monochloroaluminate hydrate.

monochloroaluminate hydrate ($3CaO.Al_2O_3.CaCl_2.10H_2O$), with a main XRD peak at 7.9 Å. The chloride ions take the place of the sulphate ions into the crystalline cell of calcium monosulphoaluminate hydrate, as the amount of calcium chloride increases. The substitution of sulphate ions for chloride ions is complete for 3% content of calcium chloride. The sulphate ions substitution is followed by a slight increase of the ettringite amount in the paste.

In our case, the acceleration introduced by calcium chloride was observed by a significant rise of XRD peaks of hydrated compounds (ettringite and portlandite) and a decrease of those of the anhydrous compounds. This difference in the hydration degree

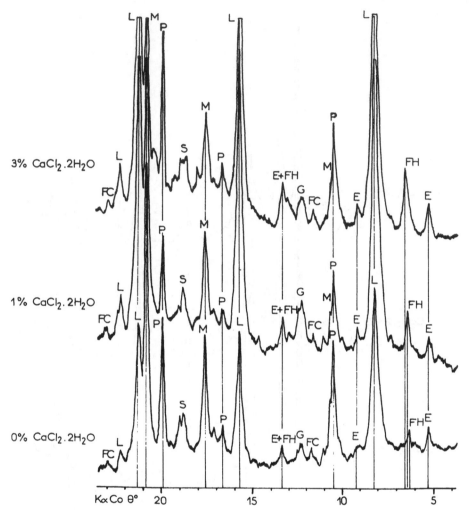

Fig. 2. X-ray diffraction diagrams of the fracture surface of 72 days old corroded steel - cement pastes composite specimens with various amounts of calcium chloride hydrate. M, magnetite; G, goethite, L, lepidocrocite; S, calcium silicates (C_2S and C_3S); P, portlandite; E, ettringite; FH, hexagonal calcium ferrite hydrate; FC, cubic calcium ferrite hydrate.

of the cement pastes in connection with the calcium chloride content is important at 3 days age but is quickly reduced with aging.

3.2 Effect of chloride ions on the microstructure of the corroded steel - cement paste ITZ

Figure 2 shows the XRD diagrams obtained on the fracture surface of 72 days old composite specimens containing various amounts of calcium chloride. We can observe the presence of significant amounts of crystallized corrosion products as magnetite, goethite and lepidocrocite and of the hydrated cement compounds, portlandite and ettringite. At the age of 72 days, both the hexagonal calcium ferrite hydrate

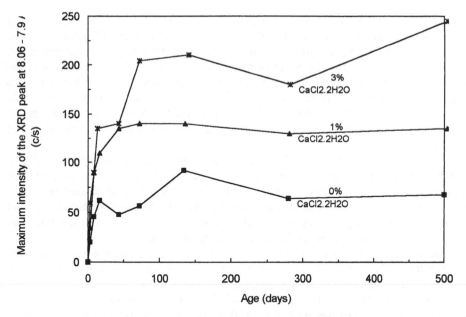

Fig. 3. Evolution with age of the maximum intensity of the X-ray diffraction peak at 8.06 - 7.9 Å of the hexagonal calcium ferrite hydrate in the ITZ of composite specimens.

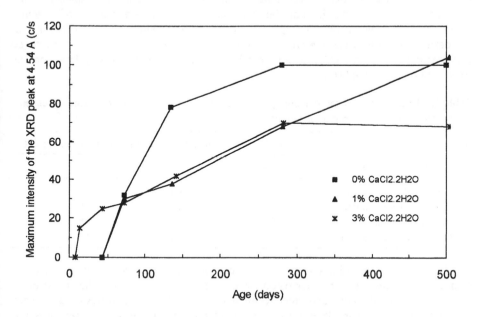

Fig. 4. Evolution with age of the maximum intensity of the X-ray diffraction peak at 4.54 Å of the cubic calcium ferrite hydrate in the ITZ of composite specimens.

($4CaO.Fe_2O_3.13H_2O$), with main diffraction peaks at 8.06, 4.03, 3.96 and 2.96 Å, and the cubic calcium ferrite hydrate ($3CaO.Fe_2O_3.6H_2O$), with main diffraction peaks at 5.20, 4.54 and 2.30 Å, are present in the fractured surface of composite specimens without calcium chloride. As the calcium chloride content in the paste increases, the intensity of the diffraction peaks of the hexagonal calcium ferrite hydrate rises significantly and its main diffraction peak at 8.06 Å diverges gradually down to 7.9 Å, showing the modification of the crystal unit cell by the intoduction of chloride ions in the crystal structure. At the same time, the diffraction peaks of the cubic calcium ferrite hydrate remain constant.

The increase in the quantity of the hexagonal calcium ferrite hydrate in the ITZ as a function of the calcium chloride content in the paste is confirmed for all the period of observation (Fig. 3). Moreover, the thickness, where the hexagonal calcium ferrite hydrate is detected by XRD, is all the more important as the calcium chloride amount increases. This thickness corresponds to the presence of the outer part of the corrosion layer in the ITZ, identified by the XRD peaks of lepidocrocite.

On the other hand, the quantity of cubic calcium ferrite hydrate and its delayed appearance in the ITZ (of one or two months) are not significantly modified by the presence of the chloride ions during the observation period (Fig. 4).

The presence of air voids in the ITZ of all the series of composites specimens leads us to think that the chloride ions, dissolved in the mixing water of the fresh cement paste, are absorbed by the pores of the corrosion layer just after casting, in the same way as dissolved calcium hydroxide is absorbed [1], [2]. If so, the quantity of chloride ions present in the corrosion layer depends directly of the absorption capacity of the corrosion layer and should not be affected by the crystallization of calcium sulphochloroaluminate hydrate or calcium monochloroaluminate hydrate in the center of the paste. But in no case, the chloride ions present in the corrosion layer have to be considered as total free chloride ions available to destroy the passivity layer on the surface of the embedded steel.

Table 1. Characteristics of the remaining layer on steel after fracture of composite specimens. Mean values for all the observation period.

Remaining layer on steel after fracture of the composite specimens	$CaCl_2.2H_2O$ amount as a % of the mass of the cement		
	0	1	3
Mean value of the thickness (μm)	100	100	110
Coefficient of variation (%)	20	25	15
Mean value of the mass (g/m^2)	185	180	195
Coefficient of variation (%)	15	20	10

Indeed, the composite specimens with calcium chloride addition do not show a significant increase of the corrosion layer nor a reduction of the mechanical characteristics of the corroded steel -cement paste bond up to 500 days. The mean thickness of the remaining layer on steel after fracture of composite specimens is about 100 μm (comparable with that of the initial corrosion layer) and does not vary significantly with the age of the paste or with the calcium chloride content (Table 1). Only, the roughness of the fracture surface increases with the age.

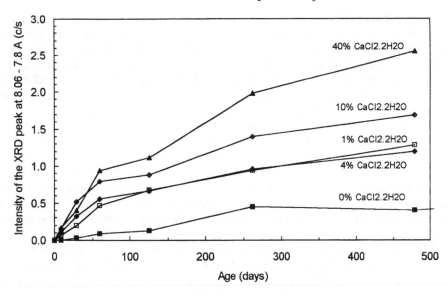

Fig. 5. Evolution with age of the intensity of the X-ray diffraction peak at 8.06 - 7.8 Å of the hexagonal calcium ferrite hydrate and calcium chloroferrite hydrate solid solution in the goethite - portlandite - calcium chloride mixtures.

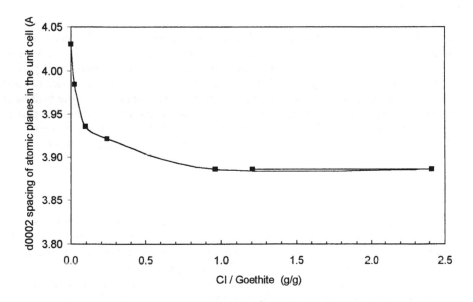

Fig. 6. Deviation of the main X-ray diffraction peak of the hexagonal calcium ferrite hydrate as a function of the chloride content in goethite - portlandite - calcium chloride mixtures.

3.3 Effect of chloride ions on the goethite - portlandite mixtures

In order to verify the results obtained on the composite specimens, pure goethite - portlandite mixtures were prepared using a goethite - portlandite weight ratio of 1/5 and a water - portlandite weight ratio of 1/1. Various amounts of pure $CaCl_2.2H_2O$ were added in the mixture : 0, 1, 4, 10, 40 and 100% of the weight of portlandite. The mixtures were cured at a temperature of $20 \pm 2°C$, protected from evaporation and carbonation and observed by XRD from 3 days age up to 500 days age.

Hexagonal calcium ferrite hydrate is detected in the mixture without calcium chloride at 3 days and its quantity increases with age (Fig. 5). When chloride ions are present in the mixture the quantity of calcium ferrite hydrate increases quickly. Its main XRD peaks deviate all the more as the calcium chloride amount is high. Closer XRD measures show (Fig. 6) that the characteristics of the unit cell of hexagonal calcium ferrite hydrate crystals with main diffraction peaks at 8.06, 4.03, 3.96 and 2.96 Å are gradually modified by chloride ions. Due to high chloride concentration in the mixture, they reach those of the hexagonal calcium chloroferrite hydrate ($3CaO.Fe_2O_3.CaCl_2.10H_2O$) with main XRD peaks at 7.8, 3.89, 3.83 and 2.93 Å.

These results confirm those observed in the ITZ of corroded steel - cement paste composite specimens. Moreover, they allow us to estimate from the deviation of hexagonal calcium ferrite hydrate XRD peaks (Fig. 2) that the weight concentration of chloride ions in the ITZ versus reactive corrosion products is about 1% when 3% of calcium chloride dihydrate are added into the cement paste.

4 Conclusion

Our study shows that the chloride ions in the cement paste significantly accelerate the crystallization of the hexagonal calcium ferrite hydrate in the outer part of the corrosion layer of the embedded reinforcement in the cement paste. Chloride ions participate in the crystal structure of hexagonal calcium ferrite hydrate and a complete series of solid solutions between hexagonal calcium ferrite hydrate and calcium chloroferrite hydrate is formed as a function of the chloride ions concentration. Therefore, an important part of free chloride ions in the ITZ is fixed in the new crystalline products, consequently reducing the risk of chloride induced corrosion for the embedded reinforcement.

5 References

1. Gallias, J. L (1998) Microstructure of the interfacial transition zone around corroded reinforcement, Second RILEM International Conference on the Interfacial Transition, Zone in Cementitious Materials, Edited by A. Katz, A. Bentur, M. Alexander, G. Arliguie, NBRI, Haifa, pp. 153-160.
2. Gallias, J. L., (1992) Etude des caractéristiques physiques et chimiques de la liaison acier corrodé - pâte de ciment, Thèse de Doctorat d'Etat, UPS, Toulouse.

16 MICROSTRUCTURE OF THE INTERFACIAL TRANSITION ZONE AROUND CORRODED REINFORCEMENT

J.L. GALLIAS
Laboratoire Matériaux et Sciences des Constructions, Université de Cergy-Pontoise, France

Abstract
Before being embedded in concrete, reinforcement may have various levels of surface corrosion due to the production process and to weathering. The corrosion layer over the reinforcement determines the characteristics of the interfacial transition zone (ITZ) with the concrete and, consequently, the quality of the bond.

Our study concerns the microstructural characterization of the corroded steel - cement paste ITZ using scanning electron microscopy, energy dispersion spectroscopy, and X ray diffraction. The corrosion layer over the reinforcement combines a physical action before setting and a chemical action during cement hydration. First, high open porosity of the corrosion layer absorbs water of the fresh paste and releases air, creating voids in the ITZ. Second, some of the steel corrosion products react with the portlandite of cement and produce hexagonal and cubic calcium ferrite hydrate.
Keywords: Interfacial transition zone, corroded reinforcement, corrosion layer, cement paste, microstructure, calcium ferrite hydrate.

1 Introduction

A builders' oral tradition says that « concrete bonds better with a corroded reinforcement than with a uncorroded one ». In fact, before concrete casting, reinforcement presents a less or more important corrosion layer over the surface of the metal. Steel bars production process and weathering during storage in building sites are the principal causes of this corrosion. The characteristics of the corrosion layer vary as the case may be. High temperature air corrosion of the steel produces a brown-black colored layer, slightly porous, bonded to the metal surface, containing crystallized iron oxides as magnetite (Fe_3O_4) hematite (Fe_2O_3) and wustite (FeO) [1]. Wet outdoors

The Interfacial Transition Zone in Cementitious Composites, edited by A. Katz, A. Bentur, M. Alexander and G. Arliguie. Published in 1998 by E & FN Spon, 11 New Fetter Lane, London EC4P 4EE, UK, ISBN: 0 419 24310 0

corrosion of steel leads to a double layer, the outer part of which is yellow-brown, very porous, very fragile and containing a mixture of colloidal iron hydroxides as $Fe(OH)_2$ and $Fe(OH)_3$ and semi-crystallized iron oxyhydroxides as lepidocrocite ($\gamma FeOOH$) and goethite ($\alpha FeOOH$). The inner part of the corrosion layer is brown-black, and contains a mixture of magnetite and goethite. The structure of the inner part of the layer is less porous and fragile than that of the outer part, but the porosity of this part of the corrosion layer is significantly higher than that of the high temperature corrosion layer [2] [3].

Our study concerns the case of the wet outdoors corrosion of the reinforcement because it seems to us that the fragility of the corrosion layer is at variance with the builders' oral tradition.

2 Materials and methods

2.1 A simple experimental model
In order to simplify the numerous factors of concrete composition influencing the characteristics of the interfacial transition zone (ITZ) around corroded reinforcement, we used neat cement paste instead of concrete for the study. The paste was prepared with an ordinary Portland cement (CPA 55R according to NF P 15 301 French standard) and with a water-cement ratio of 0.40.

A typical mild steel reinforcing bar of 20 mm diameter was used for the tests, cut in small cylindrical pieces of 15 mm height. The circular surface of the pieces was burnished up to 5 μm abrasive. The steel pieces were exposed for fourteen months to outdoors conditions without protection until a significant growth of the double corrosion layer.

At the end of the exposure period and after one day of room curing, one of the circular surfaces of the steel pieces was embedded in the cement paste. The other circular surface of the steel remained free and served to characterize the corrosion layer. The obtained corroded steel - cement paste composite specimens were cured at a relative humidity of 95 ± 5% and at a temperature of 20 ± 2°C.

The separation of the composite specimens for observation was carried out by shear stress acting tangentially to the steel surface. In these conditions, the fracture normally occurs into the ITZ. In some cases, the composite specimens were sawed perpendicularly to the steel surface after they had been embedded in a epoxy resin. The sawed surface was dry polished up to 5 μm abrasive before observation.

2.2 Characterization of composition and structure
The corroded surface of the steel pieces, the ITZ fracture surfaces of the composite specimens and the sawed surfaces of the composite specimens were observed using optical microscopy, scanning electron microscopy (SEM), energy dispersion spectroscopy (EDS) and X ray diffraction (XRD). The observations were carried out from 3 days age up to 500 days age of the cement paste.

The composition of the ITZ of the broken composite specimens was characterized using XRD on successive surfaces obtained by dry polishing on 5 μm abrasive in parallel to the fracture direction, up to the uncorroded steel surface and up to the center of the cement paste. The measure of the removed mass and of the thickness variations at any step of the abrasion process, enabled to determine the thickness and the mean apparent density of the whole ITZ in the composite specimens.

3 Results and discussion

3.1 Characterization of the corrosion layer
After fourteen months of outdoors exposure the steel presents a very irregular, rough, fine granulated layer of corrosion products. Observations by optical microscopy on sawed specimens show a clear difference between the outer part of the layer (yellow-brown colored) covering about 90% of the free surface and the inner part (black colored) covering completely the steel surface. XRD on yellow-brown and on black part of the corrosion layer confirms the above mentioned difference in composition.

The thickness of the double corrosion layer, measured by optical microscopy on sawed specimens, varies strongly between 30 and 200 μm with a mean value about 120 μm. The thickness of each part of the layer varies between 0 and 60 μm for the outer part and 20 to 100 μm for the inner part.

The mean value of the apparent dry and saturated density of the double corrosion layer was about 1450 kg/m^3 and 2400 kg/m^3 respectively. The mean value of the specific density of the corrosion products (measured on <100 μm particles) was about 3250 kg/m^3. Therefore, the porosity of the double corrosion layer is estimated about 55% and completely permeable to water.

3.2 Microstructure of the corroded steel - cement paste ITZ
The Figure 1 shows a succession of XRD diagrams obtained on parallel to the fracture surface sections between the steel surface and the center of the paste of an 8 days old composite specimen. We can observe that :

- The fracture of the composite specimen occurs in the outer part of the corrosion layer, where the most intense diffraction peaks of lepidocrocite are observed, and at the limit of the presence of the cement paste, identified by the diffraction peaks of anhydrous calcium silicates (C_3S and C_2S). Optical microscopy on the fracture surfaces show, indeed, a mosaic of yellow and black agglomerates of steel corrosion products and of translucent crystalline aggregates of cement paste compounds.
- The most significant modification of the composition of the corrosion layer is the crystallization in its outer part of some quantity of portlandite ($Ca(OH)_2$), ettringite ($3CaO.Al_2O_3.3CaSO_4.31H_2O$) and of a new crystalline compound identified as hexagonal calcium ferrite hydrate ($4CaO.Fe_2O_3.13H_2O$) with main diffraction peaks at 8.06, 4.03, 3.96 and 2.96 Å, isomorphous to the hexagonal calcium aluminate hydrate ($4CaO.Al_2O_3.13H_2O$). The outer part of the corrosion layer could be identified by the presence of strong amounts of lepidocrocite.

With the ageing of the cement paste, the diffraction peaks of the calcium ferrite hydrate become stronger, those of the lepidocrocite, goethite and portlandite decrease progressively and those of magnetite remain constant. After two months, another crystalline compound appears simultaneously with the hexagonal calcium ferrite hydrate in the ITZ, with main diffraction peaks at 5.20, 4.54 and 2.30 Å. It is identified as a cubic calcium ferrite hydrate ($3CaO.Fe_2O_3.6H_2O$) isomorphous to the cubic calcium

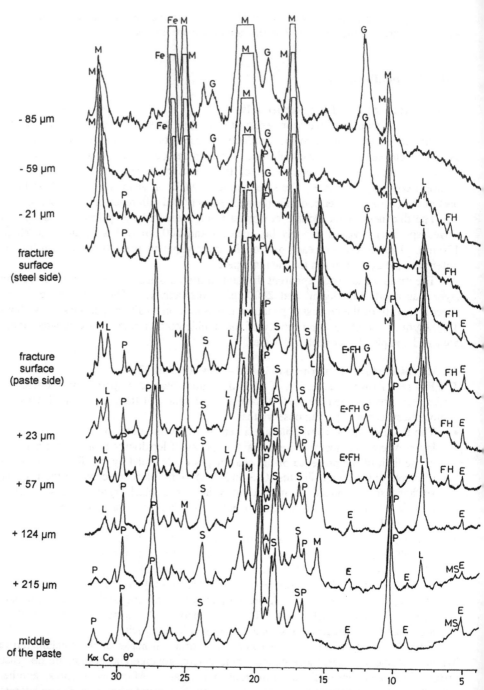

Fig. 1. X-ray diffraction diagrams within the ITZ between corroded steel and cement paste at the age of 8 days. M, magnetite; L, lepidocrocite; Fe, iron; S, calcium silicates (C_2S and C_3S); A, calcium aluminate (C_3A); P, portlandite; E, ettringite; MS, calcium monosulphoaluminate 12-hydrate; FH, hexagonal calcium ferrite hydrate ($4CaO.Fe_2O_3.13H_2O$).

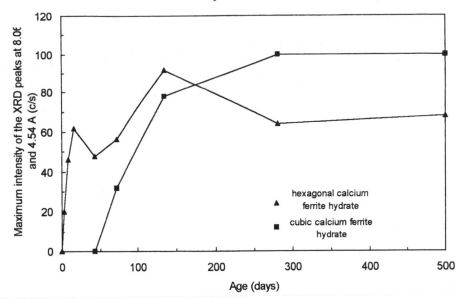

Fig. 2. Evolution with age of the maximum intensity of the main X-ray diffraction peaks of the hexagonal and the cubic calcium ferrite hydrates in the ITZ of composite specimens.

Fig. 3. Hexagonal plates formed into the outer part of the corrosion layer observed by SEM after 2 months contact with the cement paste (left-hand picture) and after 12 months contact with the cement paste (right-hand picture).

aluminate hydrate ($3CaO.Al_2O_3.6H_2O$). After four months, the diffraction peaks of the hexagonal calcium ferrite hydrate decrease slightly and stabilize up to 500 days. Those of the cubic calcium ferrite hydrate increase continually during this period. Figure 2 shows the evolution with age of the maximum intensity of the peaks of both calcium ferrite hydrates identified into the ITZ by XRD.

The SEM observations of fracture surfaces at two and twelve months shows the presence of a great number of hexagonal crystals in the middle of agglomerates of corrosion compounds (Fig. 3). The EDS analysis confirms the ferrous composition of the hexagonal crystals (Table 1) and their molar $CaO/(Fe_2O_3 + Al_2O_3)$ proportioning equal to 4.0 ± 0.7. Although, the SEM observations did not reveal cubic calcium ferrite hydrates which were detected by XRD.

Table 1. Mean EDS composition of the hexagonal crystals observed by SEM

Element	Mass %
CaO	41.2
Fe_2O_3	27.5
Al_2O_3	1.2
SiO_2	0.4
SO_3	0.3
K_2O	0.7
Na_2O	<0.1

These results allow us to conclude that steel corrosion products, particularly lepidocrocite, react with portlandite to form hexagonal and cubic calcium ferrite hydrates into the outer part of the steel corrosion layer. However, this conclusion raises the question of the presence of a significant quantity of portlandite in the corrosion layer without simultaneous presence of calcium silicates and calcium aluminate of the cement paste.

To answer this question, SEM observations and EDS mapping of the distribution of the principal elements have be made on the ITZ of the sawed composite specimen, aged of six months. We can see on Figure 4 a typical view of the ITZ with the steel in the left upper part of the pictures, identified by a very high iron (Fe) concentration, the cement paste in the right lower part of the pictures, identified by a high concentration of silicon (Si), aluminum (Al) and calcium (Ca). Between uncorroded steel and cement paste is the corrosion layer with high iron and calcium concentration. Silicon and aluminum do not penetrate significantly into the corrosion layer. A closer observation of the corrosion layer show that iron and calcium do not take the same places. Calcium fills the voids of the corrosion layer.

We can also observe that there is a big void of about 30 µm thickness is present between the corrosion layer and the cement paste, not filled by any of the principal elements present in the ITZ. Optical microscopy confirms that it is an ellipsoidal air bubble of 350 µm diameter, flatten out between the corrosion layer and the cement paste. Many other air bubbles have been detected along the limit of the corrosion layer and the cement paste.

The best explanation for all of these observations is the capillary absorption of the water of the fresh cement paste by the pores of the corrosion layer just after casting. It should be indicated that the corroded steel was cured for one day in room conditions

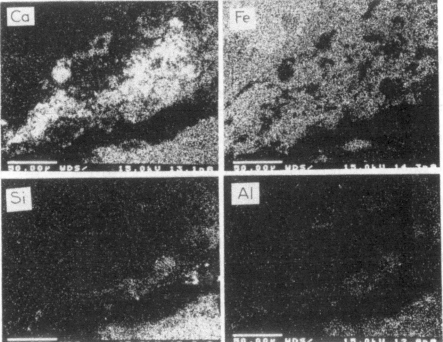

Fig. 4. SEM observation of the corroded steel - cement paste ITZ and mapping of the distribution of the principal elements : Fe, iron; Ca, calcium; Si, silicon; Al, aluminum.

before casting. Therefore, the corrosion layer was unsaturated and able to absorb water. More, the water of the fresh cement paste before setting is supersaturated with calcium hydroxide. Thus, the capillary absorption brings with water significant quantities of portlandite into the corrosion layer.

XRD diagrams in the ITZ confirm that the thickness of the cement paste affected by the water absorption, is about three times greater than the thickness of the corrosion layer. Within this thickness the hydration of the cement compounds is significantly lower than that in the center of the paste [4].

On the other hand, the water absorption of the corrosion layer ejects the included air which forms the air bubbles between the corrosion layer and the cement paste. In these conditions, a new question is presented :   Air bubbles or crystallized new compounds, which of the two determines the quality of the corroded steel - cement paste bond?

The measure of the thickness of the remaining layer on the steel after fracture of the composite specimens brings some answer to this question.   In fact, the mean thickness of the remaining layer is about 100 μm, comparable with that of the initial corrosion layer, and does not vary significantly with the age of the paste.   But, the roughness of the fracture surface increases with the age.

We can conclude that the presence of large air bubbles in the ITZ determines the place where the fracture occurs.   The new crystalline compounds formed in the ITZ, cubic and hexagonal calcium ferrite hydrates, could reinforce the corroded steel - cement paste bond only in the places where the air bubbles are not present, increasing the strength of the bond with the age.

4 Conclusion

The study of the microstructure of the corroded steel - cement paste ITZ shows that the corrosion layer on the steel combines a physical action before setting and a chemical action during cement hydration. First, high open porosity of the corrosion layer absorbs water of the fresh paste and releases air, creating air bubbles in the limit of the corrosion layer and the cement paste. Second, the steel corrosion products of the outer part of the corrosion layer (iron hydroxides and oxyhydroxides) react with the absorbed calcium hydroxide of the fresh cement paste and produce hexagonal calcium ferrite hydrate. With aging cubic calcium ferrite hydrate is formed partially in substitution for the hexagonal calcium ferrite hydrate.

The presence of the air bubbles in the ITZ probably determines the place where the fracture occurs. Therefore the mechanical characteristics of the ITZ could be improved all the more so as the corrosion layer is thinner and as its overall porosity is limited. The exposure of the reinforcement in wet outdoors conditions should be limited at few weeks. On the other hand, the crystallization of new crystalline compounds in the ITZ improves the strength of the bond. The quantity of the new compounds depends on the quantity of iron hydroxides and oxyhydroxides in the corrosion layer, and consequently on the thickness of the outer part of the corrosion layer.

5 References

1. Evans, U.R. (1968) The corrosion and oxidation of metals E. Arnold Ltd, London.
2. Gallias, J. L., Arliguie, G. and Grandet, J. (1991) Evolution de la couche d'oxydation d'une armature dans le béton, Second CANMET/ACI International Conference on Durability of Concrete, ACI, Detroit.
3. Gallias, J. L., (1992) Etude des caractéristiques physiques et chimiques de la liaison acier corrodé - pâte de ciment, Thèse de Doctorat d'Etat, UPS, Toulouse.
4. Gallias, J. L. and Grandet, J. (1992) Physicochemical reactions between corroded steel and cement paste, Second International Symposium on Phase Interaction in Composite Materials, COMP'88, Patras.

17 EFFECTS OF INTERFACIAL TRANSITION ZONE AROUND REINFORCEMENT ON ITS CORROSION IN MORTARS WITH AND WITHOUT A REACTIVE AGGREGATE

M. KAWAMURA
Department of Civil Engineering, Kanazawa University, Kanazawa, Ishikawa 920, Japan
D. SINGHAL
Regional Engineering College, Jalandhar, India
Y. TSUJI
Kajima Corporation, Tokyo, Japan

Abstract
This paper aims at comparing the differences in the degree of corrosion between steel bars embedded in mortars with and without reactive aggregate immersed in sea water and 0.51 M NaCl solution. The occurrence of ASR increased Cl^- ion concentration in the pore solution in the mortars, probably due to the transportation of Cl^- ions through cracks, and reduced the OH^- ion concentration. However, the corrosion of steel in the mortars was suppressed by the formation of massive alkali-silica gel layers around steel bars.

Keywords: alkali silica reaction; corrosion of reinforcement; sea water; interfacial transition zone; alkali silica gel layer; calcium hydroxide

1 Introduction

There are a number of concrete structures damaged by both ASR and corrosion of reinforcement under saline environments in Japan. It has been evidenced by some investigations on existing ASR damaged concrete structures and many laboratory tests that the intrusion of NaCl into concrete accelerated expansion of concrete due to ASR [1][2][3]. Under saline environments, the intrusion of NaCl through cracks in the damaged concretes appears to make reinforcement susceptible to corrosion. In fact, Vassie [4] reported that ASR cracking increased absorption of chloride and therefore tended to exacerbate corrosion of reinforcement. However, there a report suggesting that the occurrence of ASR did not necessarily exacerbate corrosion of reinforcement [5]. Thus, it is uncertain how ASR affects the corrosion of steel reinforcement in concrete at present.

The presence of calcium hydroxide surrounding steel bar in concrete, and the adhesion between steel and concrete are important factors for preventing steel from corroding [6][7]. ASR in concrete may influence corrosion of the steel through the alteration of microstructure in the interfacial transition zone around steel bars. Kawamura et al. [8] revealed that the increase in corrosion degree of steel due to ASR in mortars with NaCl added at the mixing stage was attributable to the increase in the Cl^-/OH^- ratio in the pore solution and some changes in the microstructure of mortar phase due to ASR. However, the details of microstructure of interfacial transition zone around steel bar in mortar containing reactive aggregate have not been revealed in relation to the corrosion of steel. This paper aims at comparing differences in the degree of corrosion between steel bars embedded in mortars with and without reactive aggregate immersed in 0.51 M NaCl solution and in sea water. The characteristics of

The Interfacial Transition Zone in Cementitious Composites, edited by A. Katz, A. Bentur, M. Alexander and G. Arliguie. Published in 1998 by E & FN Spon, 11 New Fetter Lane, London EC4P 4EE, UK, ISBN: 0 419 24310 0

microstructure of the interfacial zone around steel bars were examined by SEM-EDS analysis, being related to the degree of their corrosion.

2   Experimental outline

2.1   Materials
Blue Circle calcined flint (C.F.) was used as a reactive aggregate. The reactive aggregate had a size fraction of 2.5 mm to 0.6 mm. Its dissolved silica (Sc) and reduction in alkalinity (Rc) in ASTM chemical test were 1063 and 70 m mol/l, respectively. The Japanese standard sand was used as a non-reactive aggregate. Two ordinary portland cements with an equivalent percentage of Na_2O of 0.67 and 1.12, which are called, in this paper, as medium and high alkali cement, respectively, were used. Their chemical compositions are given in Table 1.

Table 1   Chemical composition of cements (%)

	CaO	SiO_2	Al_2O_3	Fe_2O_3	SO_3	MgO	Na_2O	K_2O	TiO_2	P_2O_5	MnO	Ig.Loss
High Alkali	62.5	20.1	4.7	3.2	3.3	2.5	0.41	1.08	0.26	0.06	0.06	1.2
Medium Alkali	63.6	22.0	5.2	2.7	1.9	1.5	0.40	0.41	---	---	---	1.3

2.2   Production of mortar specimens and expansion test
Mortars were prepared with an aggregate : cement ratio of 0.75 and water : cement ratio of 0.4. Two types of mortars with different reactive aggregate replacement for standard sand of 0.18 and 0.60 (reactive aggregate : total aggregate ratio by mass) and reactive aggregate-free mortar as the reference mortar were produced. A mild steel and a counter stainless steel electrode with diameter 10 mm, were embedded at a depth of 15 mm from the exposed surfaces in beam specimens, 60 mm by 100 mm by 160 mm. All the faces except a 100 mm by 160 mm face in the beam mortar specimens were coated with paraffin wax for ions to intrude into mortar from only the uncoated face. The counter stainless steel electrode was used for monitoring the electrochemical behavior of mild steel electrodes, although the results of electrochemical measurements are not referred to in this paper. After curing in a moist environment for 56 days at 20°C, mortar specimens were immersed in an artificial sea water and 0.51 M NaCl solution at 20°C. Measurements of length changes with time for mortar specimens were made on the exposed face of 100 mm by 160 mm.

2.3   Evaluation of the corrosion degree of steel bars
At 490 days after immersion in sea water and NaCl solution, steel bars were carefully taken out from mortar specimens so as not to disturb the interface between steel bar and mortar by splitting the specimens. Only some parts of the surfaces of steel bars were corroded. The corrosion degree was evaluated by the corroded area percentage, which was defined as the percentage of corroded areas to the total surface area of a steel bar.

2.4   SEM examination on interfacial mortar surfaces
Mortar pieces with undisturbed surfaces which were originally in contact with steel bar, were taken out from broken specimens. The mortar surfaces were then sputter coated with a layer of palladium.

2.5   Pore solution analysis
All the faces of mortar cylinders, ϕ40 by 100 mm, except an end face were coated with paraffin wax. The mortars with reactive aggregate were made at a reactive aggregate replacement for standard sand of 0.18. They were cured in a moist environment for 56 days at 20°C, and then immersed in sea water and NaCl solution for the prescribed periods of 30, 90, 180 and 360 days. Various ions were allowed to intrude into mortar cylinders only from the exposed end during immersion. After each immersion period,

the cylinders were removed from sea water and solution, and a group of 10 specimens was cut into 10 mm thick discs from the exposed end. Ten discs from the same depth were grouped together, and then transferred to a pore solution expression device.

3   Results and discussion

3.1   Pore solution composition and expansion of mortar

As shown in Figs. 1 and 2, the Cl⁻/OH⁻ ratio in the pore solutions for the portions to a depth of 10 mm from the exposed end increased with time. Naturally, the Cl⁻/OH⁻ ratio for mortars with reactive aggregate was raised by the consumption of OH⁻ ions through the progressive alkali silica reaction, except that, the Cl⁻/OH⁻ ratios in reactive aggregate-free mortars with a high alkali cement immersed in sea water for 180 and 360 days were about twice as high as those in corresponding reactive aggregate-containing mortars. A comparison in the Cl⁻/OH⁻ ratio between mortars immersed in sea water and NaCl solution indicates that the Cl⁻/OH⁻ ratio for mortars in NaCl solution was far greater than in sea water at 360 days.

Fig. 3 shows a BSE micrograph of a polished section parallel to a cylinder axis, which are close to the exposed end, in a mortar specimen at 490 days after immersion in sea water aerated during the tests. It is found from Fig. 3 that dense bands were formed over the exposed end in contact with sea water. The bands are found to be about 100 μm in width. SEM-EDS and XRD analyses evidenced that the outmost layer of the dense bands mostly consisted of aragonite and the inner parts of them brucite. The formation of such dense layers might hinder the intrusion of the Cl⁻ ions into mortar cylinders and the leakage of OH⁻ ions from them.

The expansions at about 500 days for all the mortars with reactive aggregate ranged from 0.45 to 0.70 %. The exposed faces of beam specimens were severely cracked.

3.3   Corrosion degree of steel bars in mortars

The external appearance of steel bars taken from the mortar beam specimens made with a medium alkali cement immersed in sea water for 490 days are shown in Fig. 4. The surfaces of steel bars in all of the reactive aggregate-containing mortars were found to be covered with a white substance, but not in reactive aggregate-free mortars. The white substance on steel bars will be discussed later. Fig. 4 also shows that all the steel bars corroded around their ends. The corrosion around the ends of steel bars may be mainly due to the intrusion of Cl⁻ ions through the crevices formed by incomplete bonds between the polymer plate and mortar specimen. Therefore, the corrosion areas around ends were not counted in calculations of corroded area percentage.

Fig. 5 shows the corroded area percentages for steel bars embedded in mortars produced with a medium alkali cement. As shown in Fig. 5, steel bars corroded only in the reactive aggregate-free mortars immersed in sea water, but only slight corrosion occurred on steel bars in the reactive aggregate-containing mortars. As shown in Fig.2 (a), the Cl⁻/OH⁻ ratio up to the depth of 10 mm in mortars made by the use of a medium alkali cement and reactive aggregate at 360 days was considerably greater than that for reactive aggregate-free mortars. Thus, from a point of view of the pore solution composition, the lower corrosion degree found in the steel bars in reactive aggregate-containing mortars was unexpected.

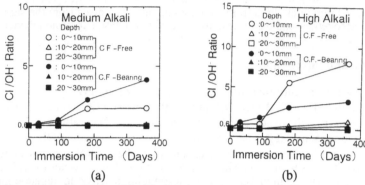

Fig. 1 Changes in Cl⁻/OH⁻ ratio with time in mortars immersed in sea water;
(a) medium alkali cement mortar, (b) high alkali cement mortar.

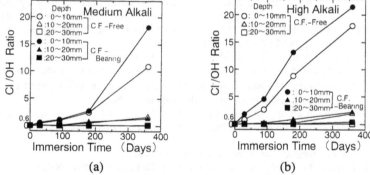

Fig. 2 Changes in Cl⁻/OH⁻ ratio with time in mortars immersed in 0.51 M NaCl
solution; (a) medium alkali cement mortar, (b) high alkali cement mortar.

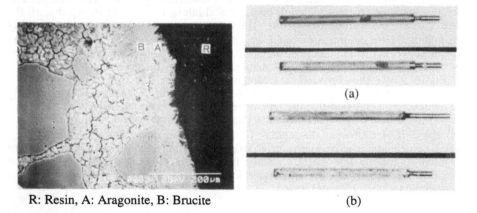

R: Resin, A: Aragonite, B: Brucite

Fig. 3 BSE micrograph of a polished section
parallel to a cylinder axis.

Fig.4 Surfaces of steel bars embedded
in medium alkali cement mortars in
sea water;(a)C.F-free,(b)C.F.-
bearing mortar.

It was also found from megascopic observations on the steel bars that amounts of a
white substance covering over the surfaces of steel bars in high alkali cement mortars

were greater than in medium alkali cement mortars. The corroded area percentages for steel bars in high alkali cement mortars are presented in Fig. 6. Greater areas on the surfaces of steel bars in high alkali cement mortars are found to be corroded compared to the steel bars in medium alkali cement mortars. This result does not contradict with the fact that the Cl^-/OH^- ratios in high alkali cement mortars were considerably greater than those in medium alkali cement mortars (Figs. 1 and 2).

Replacement of greater amounts of reactive aggregate resulted in lower degree of corrosion in steel bars in mortars made with a high alkali cement except that the steel bars in reactive aggregate-free mortars in sea water showed lower corroded area percentage than those in reactive aggregate-containing mortars with a replacement of 0.18. As a whole, lower corrosion degree in reactive aggregate-containing mortars appeared to be due to the formation of white substance layers over the surfaces of steel bars.

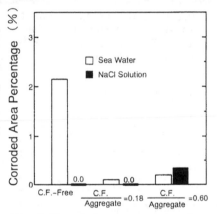

Fig. 5 Corroded area percentage for steel bars embedded in medium alkali cement mortars.

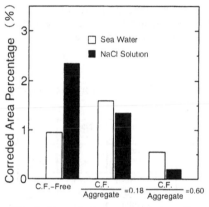

Fig.6 Corroded area percentage for steel bars in a high alkali cement mortars.

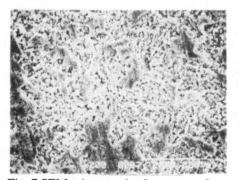

Fig. 7 SEM micrograph of mortar surfaces. in the C.F.-free mortar specimen.

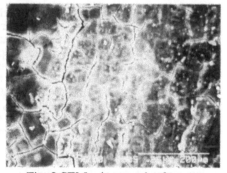

Fig. 8 SEM micrograph of mortar surfaces in C.F.-containing mortar specimen.

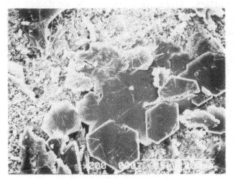

Fig. 9 SEM micrograph of portions torn off in the massive alkali silica gel layers in a medium alkali cement mortar.

Fig. 10 SEM micrograph of mortar surfaces in contact with steel bar in a C.F.-containing mortar made with a high alkali cement.

Table 2.   EDS analyses for the massive gel layers on steel bar (mol%).

Type of Mortar	Solution	Na_2O	Al_2O_3	SiO_2	SO_3	Cl	K_2O	CaO	Fe_2O_3
Medium Alkali Cement	Sea Water	6.33 (1.7)	3.42 (0.6)	38.78 (2.5)	0.95 (0.4)	0.23 (0.1)	0.65 (0.1)	48.82 (4.1)	0.81 (0.3)
	0.51M NaCl Solution	6.13 (1.8)	2.04 (0.9)	26.94 (6.9)	0.75 (0.4)	0.27 (0.2)	0.49 (0.2)	62.65 (7.4)	0.63 (0.5)
High Alkali Cement	Sea Water	12.44 (2.4)	0.49 (0.6)	45.22 (4.3)	0.99 (0.2)	0.00 (0.2)	6.39 (0.3)	34.31 (6.0)	0.15 (0.2)
	0.51M NaCl Solution	9.61 (1.1)	1.89 (0.5)	55.72 (0.8)	2.11 (0.5)	0.20 (0.1)	7.08 (0.7)	23.17 (1.2)	0.22 (0.1)

Values in parantheses are standard deviations.

3.4   Effect of steel-mortar interfacial transition zone on corrosion

SEM-EDS analyses were made on mortar surfaces in contact with steel bars in order to elucidate effects of features of steel-mortar interfacial microstructure on the corrosion of steel bars. Fig. 7 shows a SEM micrograph for the surfaces of mortar in medium alkali cement mortar beam specimens without reactive aggregate at 490 days after immersion in sea water. As a whole, the textures of the surfaces of mortar in the specimens in sea water were coarser than in those in NaCl solution. Originally, such characteristic microstructure of the interfacial zones has been shown by Bentur et al. [9] and Monteiro et al. [10]. The results of pore solution analyses (Fig. 1 (a), Fig. 2 (a)) show that various ions have reached the steel-mortar interfacial transition zone in specimens even in medium alkali cement mortars by the time of SEM observations. Therefore, the differences in the textures between mortar specimens immersed in sea water and NaCl solution may be due to the presence of ions other than Cl⁻ ions in sea water.

Fig. 8 shows a SEM micrograph for the surfaces of mortar in contact with steel bar in medium alkali cement mortar specimens containing the calcined flint at a replacement level of 0.18. A smooth appearance of the surfaces of mortar in reactive aggregate-containing mortars is greatly different from a textured appearance in reactive aggregate-free mortars (Fig. 7). The average molar compositions obtained by SEM-EDS analyses for the smooth massive layers at 5 to 9 spots are given in Table 2. The compositions in

Table 2 show that the substance having a smooth appearance was a massive alkali silica gel . Thus, a white substance seen on the surfaces of steel bars in reactive aggregate-containing mortars (Figs. 4) was found to be the massive alkali silica gel sticking to the steel surface side. It is also conjectured from the SEM observations of smooth massive gel layers that the alkali silica sols produced in mortar matrices around steel bar impregnated the steel-mortar interfacial zones. As Diamond [11] proposed, the alkali silica sol may cause calcium hydroxide in the duplex film of the interfacial zones to be dissolved, resulting in the formation of calcium-rich gels (Table 2) [12].

The surfaces of mortar were mottled by sticking of some parts of gel layers on to the surfaces of steel at splitting of specimens. Fig. 9 shows a SEM micrograph for portions partly torn off in the massive alkali silica gel layers. Large CH crystals are found to exist along with ettringite in the interfacial transition zone. Large CH crystals have been found in the interfacial zone between steel and cement paste in the presence of chloride [10][13].

Fig. 10 shows a SEM micrograph of the surfaces of mortar in contact with steel bar in high alkali cement mortar specimens containing the calcined flint at a replacement level of 0.18. The massive alkali silica gel existed in the mortar-steel interfacial zone. The results of EDS spots analyses for the gels are presented in Table 2. As given in Table 2, the sodium and potassium contents in the gels are higher than those in the gels found in the medium alkali cement mortar (Fig. 8). Far greater contents of potassium in the gels in high alkali cement mortars explicitly reflect a high potassium content of the cement used. Relatively high calcium content in the gels suggests that the calcium in the gels was enriched in the process of transport of the sols.

Page [6] pointed out the significance of calcium hydroxide surrounding the steel as a protective role of mortar in corrosion. However, the situation of the interfacial zones in reactive aggregate-containing mortars was greatly different from that in the usual mortars in the respect that the steel was in contact with the massive alkali silica gels. The pH of alkali silica gels is considered to range from 10.8 to 12.8 [11]. As previously described, the corrosion degree of steel in reactive aggregate-containing mortars was lower than in reactive aggregate-free mortars. Therefore, the depression of corrosion in reactive aggregate-containing mortars in sea water and NaCl solution may result from the formation of homogeneous alkali silica gel layers surrounding steel bars. However, the amounts of alkali silica gels produced, the alkali silica reaction rate and the extension of cracks in concrete depend on various factors. A universal concept on the effects of the alkali silica reaction on corrosion warrants further study and investigation.

4   Conclusions

(1) The Cl^-/OH^- ratios in mortars in 0.51 M NaCl solution were greater than in sea water at 360 days after immersion. The formation of dense layers of aragonite and brucite might hinder the intrusion of the Cl^- ions into mortar cylinders and the leakage of OH^- ions from them.

(2) From the pore solution composition point of view, far lower corrosion degree in steel bars in reactive aggregate-containing mortars was unexpected.

(3) Massive alkali silica gels existed in the mortar-steel interfacial zone. The calcium content in the gels may be increased in the process of transport of sols.

(4) The presence of large CH crystals in the interfacial zone was confirmed.

(5) Replacement of greater amounts of reactive aggregate for standard sand in mortars resulted in lower degree of corrosion in steel bars, except for one case. The depression of corrosion in reactive aggregate-containing mortars in sea water and NaCl solution appears to result from the formation of homogeneous alkali silica gel layers surrounding the steel bars.

5   References

1. Kawamura, M., Torii, K., Takeuchi,K. and Tanikawa, S.(1996) Long-term ASR expansion behavior of concrete cubes in outdoor exposure conditions, Proc. of the 10th Intl. Conf. on Alkali-Aggregate Reaction in Concrete, Melbourne, pp.630-36.
2. Kawamura, M.and Takeuchi, K. (1996) Alkali silica reaction and pore solution composition in mortars in sea water, Cem. & Concr, Res., Vol. 26, No.12, pp. 1809-20.
3. Kawamura, M., Takeuchi, K. and Sugiyama, A. (1996) Mechanisms of the influence of externally supplied NaCl on the expansion of mortars containing reactive aggregates, Magazine of Concrete Research,Vol.48, No.176, pp.237-48.
4. Vassie, P. R. (1993) Secondary effects of ASR on bridges-corrosion of reinforcement steel, Proc. 2nd Intl. Conf. Bridge Management, University of Surry, Guilford, pp.18-21.
5. Sanjuan,M.A.andAndrade,C.(1994) Effect of alkali-silica reaction on the corrosion of reinforcement, Supplementary Papers, Third CANMET /ACI Intl. Conf. on Durability of Concrete, Nice, pp.613-22.
6. Page,C.L.(1975) Mechanism of corrosion protection in reinforced concrete marine structure, Nature, Vol. 258, pp.514-15.
7. Yonezawa,T.(1988) Pore solution composition and chloride-induced corrosion of steel in concrete, Ph.D thesis, UMIST, p.426.
8. Kawamura, M., Takemoto, K. and Ichise, M.(1989) Influence of the alkali-silica reaction on the corrosion of steel reinforcement in concrete, Proc. 8th Intl. Conf. on Alkali-Aggregate Reaction, Kyoto, pp.115-20.
9. Bentur,A., Diamond, S. and Mindess, S.(1985) The microstructure of the steel fiber-cement interface, J. of Materials Science, Vol.20, pp.3610-20.
10. Monteiro, P.J.M., Gjorv, O.E. and Mehta, P.K.(1985) Microstructure of the steel -cement paste interface in the presence of chloride, Cem. & Concr. Res., Vol.15, No.5, pp.781-84.
11. Diamond, S.(1983) Alkali reaction in concrete-pore solution effects, Proc. 6th Int. Conf. on Alkalis in Concrete, Copenhagen, pp.155-66.
12. Scrivener, K.L.and Monteiro,P.J.(1944) The alkali-silica reaction in a monothilic opal, J. Amer.Ceram.Soc., Vol.77, No.11, pp.2849-56.
13. Diamond, S.(1986) The microstructure of cement paste in concrete, Proc. 8th Intl. Congress on the Chemistry of Cement, Vol.1, pp.122-47.

18 EFFECTS OF THE REINFORCEMENT CORROSION INDUCED BY CHLORIDES ON THE INTERFACE STEEL–CEMENT PASTE IN MORTARS

E. MENÉNDEZ, M.A. SANJUÁN and C. ANDRADE
Instituto "Eduardo Torroja" de Ciencias de la Construcción, CSIC, Madrid, Spain

Abstract
Chlorides are the main environmental agent producing damage in reinforced concrete structures via the metallic corrosion and the products spread out through the pore microstructure producing microstructural changes which affect the stability of the material leading to the failure of the structure. In order to getting deeper knowledge of this process, calcium chloride was added to the mixing water of a mortar to promote the reinforcement corrosion. After two years of exposure at 100% RH, the specimens were prepared to be studied by means of the back-scattering electron microscope. EDS analyses were also performed at the interface mortar-steel.
Keywords: BSE microscopy, Chlorides, Corrosion, Cover, Microstructure, Mortar, Reinforcement.

1 Introduction

Chloride contaminated environments could promote durability problems in concrete structures. This is due to the local disruption of the passive layer of oxides formed on the surface of the steel in alkaline medium This attack is induced by the local acidification promoted by the chlorides (1) and may eventually neutralize the alkaline nature of the pore solution and enable the dissolution of cement phases (2).

The oxides formed in the corroding areas tend to diffuse through the pores and press the concrete cover out and producing the cracking.

The main factors influencing the depasivation process are the chloride concentration, the pH of cement ($[Cl_-]/[OH_-]$ ratio), the diffusion of oxygen through the concrete which fixes the stell electrical potential, and the electrical resistivity of the concrete.

The Interfacial Transition Zone in Cementitious Composites, edited by A. Katz, A. Bentur, M. Alexander and G. Arliguie. Published in 1998 by E & FN Spon, 11 New Fetter Lane, London EC4P 4EE, UK, ISBN: 0 419 24310 0

In this study the steel/mortar interface is investigated by means of back-scattered electron imaging which has enabled to observe some features of the effect of the corrosion oxides in the microstructure of the mortar surrounding the corrosion zones.

2 Experimental

Mortar specimens of 20x 55x 80mm, with a 0.5 water-cement and 1/3 cement-sand ratios were made. 1 % of Calcium chloride has been added to the mixing water to promote the reinforcement corrosion. Mortar specimens without chlorides were also prepared for reference. The composition of the normal portland cement (opc) used is given in Table 1. The sand grading ranged between 0 and 2mm, and distilled water was used. The specimens were cast under laboratory conditions. They were removed from the moulds after 24 hours and then they were storaged at 100% at lab temperature for one year.

Table 1. Chemical composition of the normal portland cement CEM I 42,5 R (% by weight)

SiO_2	Al_2O_3	Fe_2O_3	CaO	TiO_2	MgO	SO_3	Na_2O	K_2O	L.O.I.	I.R.
20.0	5.6	2.2	64.7	0.23	0.9	3.1	0.19	0.58	1.8	0.4

2.1 Sample Preparation and Electron Microscopy
Polished sections of the mortar samples were prepared for SEM examination. After one year of storage, one half of each specimen was cut longitudinally to the rebar for microstructural examination. Sections were then cut transversally to the steel, resin impregnated, lapped with 78µm, 46µm, 25µm and 18µm grit and polished with 6µm and 1µm diamond powder to obtain a flat surface. The samples were then coated with carbon.

A JEOL JSM-5400 scanning electron microscope (SEM) (accelerating voltage of 20keV) was used equipped with secondary and back-scattered electron detectors, fitted with a Link System ISIS-Oxford energy dispersive X-ray spectrometer (EDS). Point analyses were made in the region of the steel/mortar interface, particularly to investigate the presence of chloride ions. In addition, an X-ray dot map was made of a section of the interfacial area to indicate the relative concentrations of Ca, Al, Fe and Cl. Also optical microscopy was used. The optical microscope was an stereomicroscopy NIKON SMZ-27 with a photographic system NIKON Microflex AFX-DX wiht a photographer camera NIKON FX-35 DX.

3 Results and Discussion

First optical microscopy was used to study both samples without (figure 1) and with 1% calcium chloride (figure 2). A good adherence of steel-mortar and a lack of corrosion around the steel can be observed in figure 1, while the mortar mixed with 1% calcium chloride presented a significant amount of oxides (figure 2) in the section selected for the study. These oxides were preferentially distributed in the bottom of the steel, considering the upper area of the finishing side. This is a direct consequence of the higher porosity and amount of water in this area after casting. Also, there can be observe some pits on the steel.

Figures from 3 to 11 are the backscatering electron microscope micrographs and several series of X-ray dot maps of samples without (figures 3 and 4) and with 1% calcium chloride (all the rest). A good adherence aggregate-paste and steel-mortar can be observed in figure 3. Neither oxides around the steel or microcracks in the interfaces aggregate-paste and steel-paste are observed. These observations evidenced the good state of the steel rebars embedded in the mortar mixed without calcium chloride. Looking at the dot mapping analysis performed in this area, the absence the oxides in the steel-paste interface is corroborated. Large amounts of calcium were found close to the steel, likely due to the precipitation of portlandite.

With regard to the mortar mixed with 1% calcium chloride, as mentioned great amounts of oxides were observed (figure 5). Microcracks were also observed either in the mortar or in the oxide layer. In the first step, oxides tend to refill the free spaces of the mortar such as pores and microcracks as evidenced by the pore filled with oxides, which is shown in the image. The oxides (Fe window in figure 6) are mixed with calcium and silicon from the mortar to form a paste which spread through the free paths offered by the pores. Sometimes, the pressure promoted by the oxides formation leads to microcracking in the mortar (figure 7) and also the mentioned layer of oxides mixed with calcium can be broken by stresses induced in such layer.

Figure 8 shows a pit in the steel. Chloride ions in several amounts have been detected either inside the pit and around it (figure 9); whereas calcium was also detected in the interface containing the oxides. The microstructure was affected for the oxides formation, first involving a pore refilling and second a microcracking process of the matrix.

No relationship between the chloride amount and the corrosion of the steel reinforcement was found. This observation is supported by data summarized in table 2 which are the result of X-ray point analyses taken in the points marked on figure 8.

Another example is given in figure 10, in which a dot map is presented in figure 11 and x-ray point analyses are summarized also in table 2. Several pits filled with oxides which also spread out to the mortar area are observed. There is a significant amount of chloride either in the pores or in the oxide layer and calcium remains in this oxide layer, whereas is not significant the chloride content withing the pits.

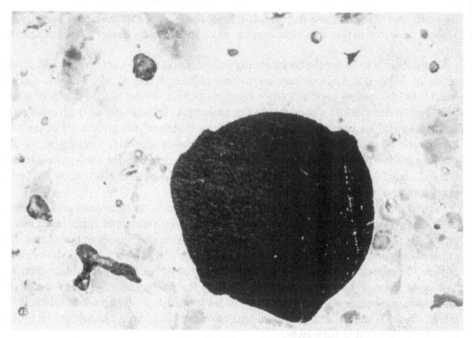

Figure 1: Optical microscopy of mortar without calcium chloride

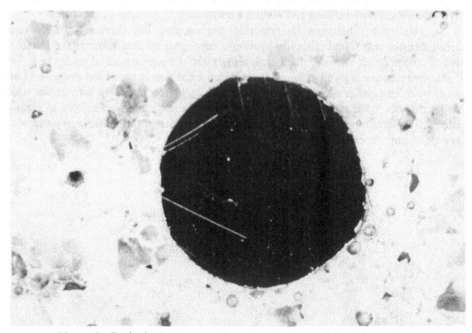

Figure 2: Optical microscopy of mortar with 1% calcium chloride

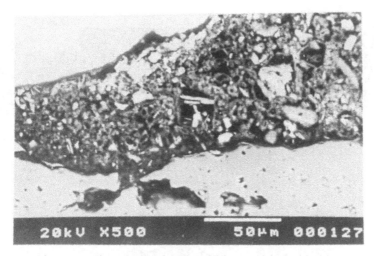

Figure 3: Micrograph of mortar without calcium chloride

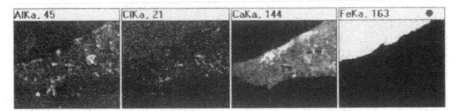

Figure 4: X-ray dot maps of samples without calcium chloride

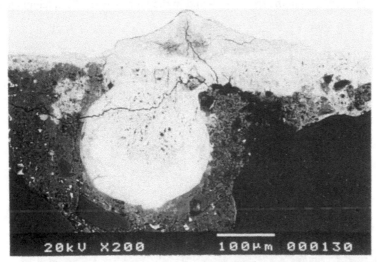

Figure 5: Micrograph of mortar with 1% calcium chloride

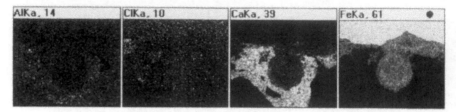

Figure 6: X-ray dot maps of samples with 1% calcium chloride

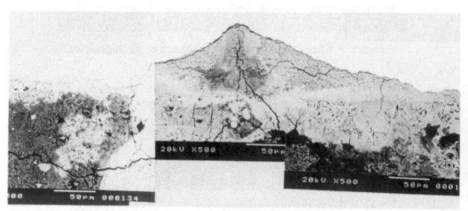

Figure 7: Microcracks in the oxide layer and in the mortar with 1% calcium chloride

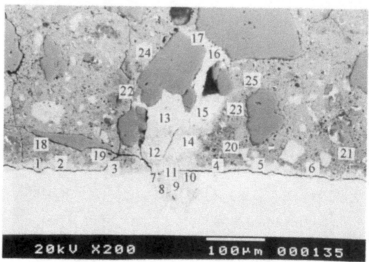

Figure 8: Micrograph of mortar with 1% calcium chloride

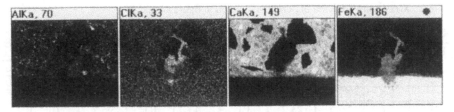

Figure 9: X-ray dot maps of samples with 1% calcium chloride

Figure 10: Micrograph of mortar with 1% calcium chloride

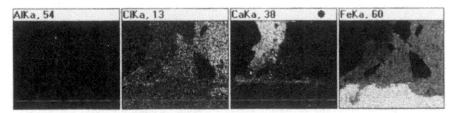

Figure 11: X-ray dot maps of samples with 1% calcium chloride

Table 2. X-ray point analyses in percentage taken in figures 8 and 10

Figure 8					Figure 10			
Point	Cl⁻	Fe₂O₃	CaO	Al₂O₃	Point	Cl⁻	Fe₂O₃	CaO	Al₂O₃

Point	Cl⁻	Fe_2O_3	CaO	Al_2O_3	Point	Cl⁻	Fe_2O_3	CaO	Al_2O_3
1	0.6	56.6	29.7	1.2	1	0.4	98.5	0	0
2	0.7	66.6	21.7	1.6	2	0.5	98	0	0
3	0.6	51.1	30.0	4.0	3	1.5	95.8	0	0
4	0.8	56.5	23.2	3.1	4	1.1	96.7	0	0
5	1.1	65.9	23.1	2.0	5	0.6	93.2	3.4	0
6	0.9	41.1	47.0	1.0	6	0.8	95.1	1.8	0
7	0.4	88.0	6.3	0.8	7	0.1	95.3	2.7	0
8	4.1	91.8	0.3	0	8	0.2	95.2	2.2	0
9	4.0	91.7	0.2	0.2	9	0.4	98.0	0.8	0
10	1.9	91.5	0.5	0.2	10	5.1	93.0	0	0
11	2.0	87.9	0.5	0.2	11	3.5	75.4	4.7	3.4
12	0.4	33.5	8.9	11.9	12	2.8	70.9	9.0	1.1
13	4.1	80.0	0.9	2.0	13	0.5	34.2	3.6	12.3
14	1.5	56.9	18.1	4.8	14	0.8	70.9	2.1	1.9
15	3.0	71.7	1.0	4.2	15	1.7	83.3	2.3	0.9
16	3.9	79.9	1.2	1.5	16	1.6	88.1	3.3	0.6
17	1.9	59.6	8.1	6.7	17	3.0	82.8	1.9	1.0
18	0.6	2.5	60.0	2.5	18	3.2	79.0	1.1	5.3
19	0.1	1.0	81.1	1.5	19	1.7	76.6	1.0	3.7
20	0.3	2.7	62.9	5.2	20	1.6	70.8	1.4	2.4
21	0.6	2.0	62.2	2.4	21	2.3	62.0	5.2	7.3
22	0.4	1.1	64.4	1.3	22	2.7	77.4	1.5	2.3
23	0.4	1.6	62.8	1.7	23	2.8	75.6	1.1	2.4
24	0.3	5.3	65.1	2.2	24	4.1	78.0	0.8	2.5
25	0.4	6.1	60.1	4.8	25	3.3	76.2	0.8	4.1
					26	3.0	75.2	0.5	1.5
					27	0.5	30.8	36.5	2.7
					28	0.5	33.3	27.5	1.5

4 Conclusion

- Samples mixed with chloride ions in the mixing water showed localised attack on the steel surface and a great amount of oxides. The corrosion oxides formed induce a microcracking which has been detected in the study. The amount of chlorides does not relate with the intensity of the attack, but is independent of the amount of oxides.
- Chlorides are always in the oxydes, and only sometimes withing the pits. The oxydes tend to spread out of the corroding zones diffusing through the pores and apparently dissolving the cement phases which disappear and form a mixture with the oxydes.

5 References

1. Page, C.L. and Treadaway, K.W.J. Aspects of the electrochemistry of steel in concrete, Nature, 297 (1982), p.109.
2. Sanjuán, M.A., Scrivener, K.L., Alonso, C. and Andrade, C. "Microstructural development of mortars containing embedded corroded steel bars". The Institute of Materials. St Anne's College de Oxford (U.K.). 26-27 Sept, 1994.
3. Constantinou, A.G., Sanjuán, M.A. and Scrivener, K.L. "Electrochemical and microstructural performance of steel reinforced carbonated and non-carbonated mortars in a saline environment". Corrosion'95. The NACE Annual Conference and Corrosion Show. Paper No. 284. 26-31 de March, 1995. pp.284/1 - 284/11.
4. Menéndez, E., Sanjuán, M.A.,and Andrade, C. "Interface corroded steel-cement paste in mortars mixed with several amounts of calcium chloride". Sixth Euroseminar on Microscopy Applied to Building Materials. Reykjavik (Iceland). 25-27 June, 1997.

19 COMPARISON OF ITZ CHARACTERISTICS AROUND GALVANIZED AND ORDINARY STEEL REBARS

F. BELAID, G. ARLIGUIE and R. FRANÇOIS
Laboratoire Matériaux et Durabilité des Constructions, UPS, Toulouse

Abstract
Minimizing the degradation due to corrosion is necessary to provide increased durability of concrete structures. In order to characterize the ITZ formed around both galvanized and uncoated steel bars, experiments are performed by using surface hardness tests, SEM observations, X-ray diffraction analysis, and water absorption tests.
Keywords: galvanized steel, corrosion, ITZ, durability.

1 Introduction

The durability of reinforced concrete structures is essentially linked to steel passivity in concrete. But in some cases, aggressive agents may contaminate the concrete cover inducing thus the corrosion process of re-bars, major degradation factor of engineering concrete structures. The use of galvanized steel is one of the solutions proposed to avoid this phenomenon when engineering concrete structures are subjected to aggressive conditions.

However, an important consideration to taken into account when using galvanized rebars is the chemical reaction that occurs between the superficial layer of galvanization and the ions of the pore solution concrete. As a result, this interaction induces a delay in the setting time in the contact surface with galvanized steel as well as a formation of an interaction compound often associated with an emission of hydrogen at the interface level.

The object of our study is to compare the ITZ formed around ordinary steel and galvanized steel. The characterization tests selected are the variation of surface hardness of cement paste in ITZ, the analysis by X-ray diffraction, the observation by SEM. The porous structure of the cement paste in ITZ is evaluated through water absorption measures.

The Interfacial Transition Zone in Cementitious Composites, edited by A. Katz, A. Bentur, M. Alexander and G. Arliguie. Published in 1998 by E & FN Spon, 11 New Fetter Lane, London EC4P 4EE, UK, ISBN: 0 419 24310 0

2 Wear by abrasion

2.1 Principle

This test, based on the Dorry test [1], is used to quantify the superficial hardness of engineering materials (aggregate, cement paste, mortar). It is based on determining mass losses after successive abrasions of one sample face with an automatic polishing machine. The implementation of the method used in this study have been done by Lucas [2].

We are working on cylindrical cement paste test pieces. The test piece, of S section, is placed in a sample holder then on an abrasive disk. The supporting force, called F, corresponds to the weight of the test piece/sample holder assembly. The path on the abrasive disk is set by a guiding plate so that the wear is uniform (Fig. 1).

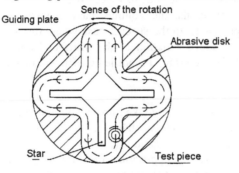

--- ----- Path of the test piece

Fig. 1. Automatic polishing machine.

This wear is expressed through the decrease of the test piece height. It increases in relation to the time of the test and the harder the material is, the lesser will be the sample shortening:

$$\Delta H = \frac{F}{S} \, t \, A$$

where A is a wear coefficient characteristic of the material depending on experimental conditions and the softer the material is, the higher coefficient will be.

The height variation of the test piece is almost impossible to determine through direct measure, considering the low values at stake. Experimentally, this height variation is obtained by weighing the test piece mass difference before and after an abrasion (Δm).

$$\Delta H = \frac{\Delta m}{\rho . S}$$

ρ is the volume mass of the material tested

S is the surface of the face abraded.

The wear coefficient by abrasion - A - is expressed as :

$$A = \frac{\Delta m}{\rho . S . t . P}$$

where P is the pressure applied on the wear surface (P=F/S)

Conventionally, A represents an abrasion wear rate. The abrasion wear rate is a characteristic of the material studied depending on the test time and on the test equipment features (supporting force, plate rotation speed, type and sizes of abrasive grains). It can be expressed in $m.s^{-1}.Pa^{-1}$, but considering its low numerical values, we shall use the $\mu m.s^{-1}.Pa^{-1}$.

When the test piece is a cement paste, it is liable to loss a significant quantity of water during abrasions. This is particularly the case for young cement pastes or cement pastes made with a high E/C ratio. To take this phenomenon into account, we determine Δm, with two similar test pieces, one being used as a test specimen to quantify the mass loss due to water evaporation and the other to perform the abrasion.

$$\Delta m = \Delta m_1 - \Delta m_2$$

Δm represents the mass difference due to abrasion only,

Δm_1 represents the mass difference of the test piece before and after abrasion,

Δm_2 represents the mass difference due to water loss.

For homogeneous materials, the volume mass is measured through hydrostatic weighing on a test specimen. In the specific case where the wear surface corresponds to an ITZ, the volume mass variation in the interfacial area cannot be determined accurately. We must use therefore, a modified coefficient of wear by abrasion defined by: $A_m = \dfrac{\Delta m}{t.P}$

This wear coefficient A_m does not take into account the volume mass of the abraded cement paste, but the error made on the wear by abrasion rate in the ITZ is less significant as if we use the A coefficient [2].

Tests are performed with the appropriate measures of precaution stated during the implementation of the method (disk running in, pressure applied ranging between 3 and 7 KPa, duration of the test limited to 60 seconds...) [2].

2.2 Type of test specimens used for this test

These tests are intended to compare the steel-cement paste and zinc-cement paste ITZ. For this, we have made mixed test specimens obtained by casting a cement paste on a metallic support (steel or zinc) whose surface state is perfectly defined.

The cement paste is made from a CPA CEM I 42,5 with different E/C ratios (0,3; 0,4; 0,5).

At the end of test periods (1, 2, 3, 7 and 28 days), the test pieces are broken. The abrasions are then performed on the cement paste face which was in contact with the metal.

Successive abrasions are performed until the A_m wear coefficient value is constant.

2.3 Result presentation

Results are presented by curves indicating the variation of A_m in relation to the distance measured from the contact with the metallic support.

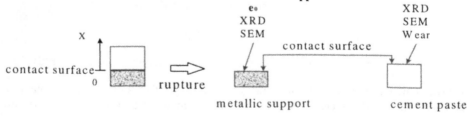

Considering that the determination of the A_m wear rate is not performed on a specific test piece surface but on a volume (surface x thickness of material removed by abrasion), the distance to the support (δ) shall be taken as the average of distances to the support before and after abrasion.

$$\delta_i = \frac{e_{i-1} + e_i}{2} \qquad \text{with } e_i = e_{i-1} + \frac{\Delta m_i}{\rho.s}$$

For the first abrasion, e_o represents the thickness of the cement paste film sticking to the support after breaking of the mixed test piece.
e_o is determined through observation of the metal side with an optical microscope.

2.4 Additional information
Analyses by X-ray diffraction are performed on steel-cement paste and zinc-cement paste contact surfaces, as well as on some surfaces showed by abrasion in test pieces. In addition, observations are performed by SEM.

2.5 Results and interpretation
At young ages, the evolution of the abrasion rate according to the distance in relation to the contact surface corresponds to the standard curves of Fig. 2.

The abrasion rate decreases and becomes stabilized when the cement paste presents the wear characteristics of the sample core.

However, the relative position of the curves depends strongly of the cement paste type and of the E/C.

After 3 days, the abrasion rate and the ITZ thickness are systematically lower in the case of the zinc contact (Fig. 2).

Microstructure observations (SEM), together with the X-ray analysis diffraction indicate that the cement paste hydration in contact with zinc is delayed. However, the calcium hydroxyzincate whose presence is revealed by X-ray diffraction analysis both on zinc support and in ITZ, seems to provide a higher hardness because the orientation of the portlandite at the ordinary steel interface make the transition area more fragile (Fig. 3).

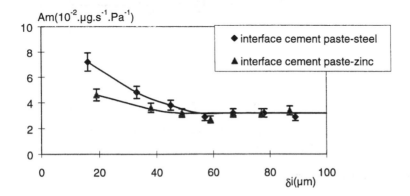

Fig. 2. Evolution of the wear coefficient of abrasion (A_m) with the distance of the support (δ_i) 3 days after casting.

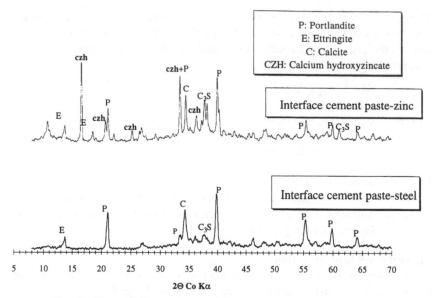

Fig. 3. X-ray diffractograms of the cement paste after rupture.

For longer periods, the cement paste hydration reduces ITZ thickness as well as the difference between the abrasion rates of the two interfaces, whatever the E/C ratio.

At 28 days, the curves representing the evolution of abrasion rates of the two ITZ are nearly similar (Fig. 4.).

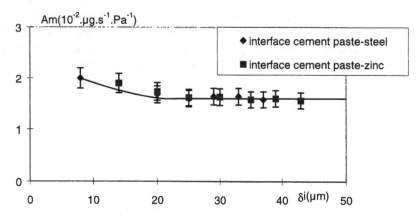

Fig. 4. Evolution of the wear coefficient of abrasion (A_m) with the distance of the support (δ_i) 28 days after casting.

3.Water absorption test

The absorption test allows to determine the ability of a porous material to absorb and transport a wetting liquid (water) through capillarity [3].   A water absorption coefficient representing the water quantity absorbed per porous material surface unit and per time unit is defined as follows:

$$a = \frac{\Delta m}{S.\sqrt{t}} \cdot \rho_{eau}$$

Δm =   water quantity absorbed by the porous material (kg)
ρ_{eau} =   water volume mass
S =   contact surface with water (m^2)
t =   suction time (s or h)
a is therefore expressed in kg.m^{-2}. s^{0,5} or in kg.m^{-2}.h^{0,5}

A priori, the water absorption coefficient should be independent of the test duration, but this is not experimentally checked [3]. It is therefore important to mention the experiment duration.

3.1 Samples and experiment

The mixed test pieces are no longer made from a cylindrical metallic support but from a steel or galvanized steel plate of 40 x 80 x 50 mm. The cement paste has a E/C ratio of 0,5.

These test pieces are broken after 19 hours or 28 days of storage. Each cement paste sample is stoved until it reaches constant mass then the side faces are coated with resin. The face which was in contact with the metallic plate is placed in contact of a free sheet of water of 1 cm height; this sheet is kept to a constant level by an overflow.

The water mass absorbed by the sample is measured through successive measurements for determined times. The total duration of the test is 24 hours.

The water absorption kinetics is not experimentally constant. It is usual to define two slopes on the curve representing the water mass absorbed by surface unit according to $\sqrt{t}$. Between 0 and 1 hour, the curve slope represents the filling of the biggest capillaries. This slope is the water absorption coefficient noted a_{1h}.

In the test sequence (between 1 hour and 24 hours), the capillary pores thinner and thinner, are involved in the water absorption process. The curve slope corresponds to the absorptivity [4] [5]. The absorptivity variation characterizes the microstructure development of the bonding phase with the hydration progression: it evolves therefore until the hydration is completed.

3.2 Results

The variation of the water mass absorbed per surface unit in contact with water is represented in accordance with the time square root (Fig. 5 and Fig. 6).

Fig. 5 corresponds to test pieces broken 19 hours after the casting of the test piece. The curves of steel-cement paste and galvanized steel-cement paste ITZ are very similar until about a quarter of an hour and then deviate progressively. This deviation is more apparent at the slope level characterizing the filling of the thinnest pores (slope between 1h and 24 h).

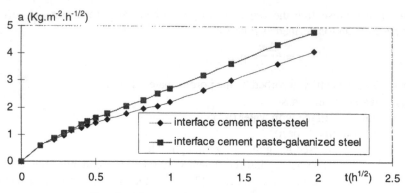

Fig. 5. Absorptivity curve at 19 hours.

For test pieces broken 28 days after the casting of the test piece, the curves are practically the same : there is no difference between the steel-cement paste ITZ and the zinc-cement paste ITZ (Fig 6).

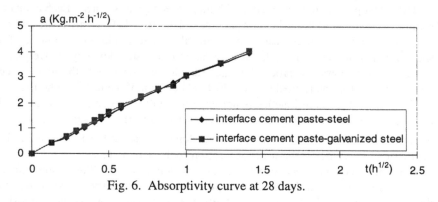

Fig. 6. Absorptivity curve at 28 days.

4 Global interpretation of results

The different measures that have been achieved show that the ITZ formed between a metal and the cement paste changes over time.

In the case of ordinary steel, the variations of the physical characteristics in the ITZ correspond to :

- a higher rate of wear by abrasion in the contact zone with steel decreasing as we go away from the metal,
- hydrates better crystallized as in the cement paste core,
- a preferential orientation of Portlandite crystals.

In the case of galvanized steel, the ITZ characteristics are fully conditioned by the interfacial reaction occurring between the superficial galvanizing layer and the cement paste. Therefore, the chemical composition of the cement used and the superficial galvanizing layer type play a determining role. The conditions used in this test series have led to the formation of calcium hydroxyzincate in the interfacial zone (Fig. 3).

This compound crystallizes in a disorderly way and induces a high quality of bonding between galvanized steel and cement paste [6]. This produces a lower rate of wear by abrasion than in the case of steel ITZ.

For young ages, water absorption curves reveal a higher degree of absorption in the zinc ITZ than in the steel ITZ. This corresponds to the gaseous emission which often accompanies the zinc-cement paste reaction. However, this absorptivity difference decreases over time which means that, in the case studied, cement hydration is not inhibited in the ITZ and allows a filling of the porous structure in this contact area.

5 Conclusion

The results obtained confirm that the ITZ formed around reinforced concrete steels presents different physical characteristics from that of the cement paste corresponding to a weaker zone. Nevertheless, the microstructure of this zone changes over time and the difference with the cement paste core diminishes without disappearing however.

As far as the interface with galvanized steel is concerned, we know that it strongly depends on the chemical reaction which is liable to occur. This test series, made with a single type of cement and galvanized steel represents therefore a specific case. However, it reveals that, to have good characteristics in this zone, a change over time must be possible which means that cement hydration must not be inhibited.

6 References

1. Kasa and Subert (1984) Hungarian experiences gained with the aggregate abrasion test, *Symposium international sur les granulats thème III*, Nice.
2. Lucas,J.P (1985) Etude, par une méthode d'usure par abrasion, de l'auréole de transition formée dans un béton de ciment portland, *Thèse de Doctorat*, Toulouse.
3. Yssorche,M.P (1995) Microstructure et durabilité des bétons à haute performance, *Thèse de Doctorat*, Toulouse.
4. Balayssac,J.P, Detriche,Ch.H, Grandet,J (1993) Intérêt de l'essai d'absorption d'eau sur la caractérisation du béton d'enrobage, *Materials and structures*, Vol 26, pp 226-230.
5. Hall,C (1989) water sorptivity of mortars and concretes: a review, *Magazine of concrete and research*, Vol 41, No147 pp51-61.
6. Arliguie, G , Grandet, J (1982), Formation de l'auréole de transition entre la pâte *de* ciment portland et le zinc, *Colloque international Rilem: Liaisons pâtes de ciments -matériaux associés*, Toulouse.

PART SIX

ITZ STRUCTURE AND PROPERTIES IN PORTLAND CEMENT SYSTEMS

20 APPETENCY AND ADHESION: ANALYSIS OF THE KINETICS OF CONTACT BETWEEN CONCRETE AND REPAIRING MORTARS

L. COURARD and A. DARIMONT
Université de Liège, Institut du Génie Civil, Laboratoire des Matériaux de Construction, Liège, Belgium

Abstract
The definition of the adhesion itself presents a duality : on the one hand, adhesion is understood to be a process through which two bodies are brought together and attached (bonded) to each other, in such a manner that external force or thermal motion is required to break the bond. The term "adhesion" (or "sticking") is usually applied in this sense in colloid research on coagulation phenomena.
On the other hand, we may examine the process of breaking a bond between bodies that are already in contact; and here, as a quantitative measure of the intensity of adhesion, we can take the force or energy required to separate the two bodies.
We studied this problem in the context of repairing mortars laid down on a concrete support; we focused our research on the behaviour of a cementitious slurry applied on two types of concrete support, characterised by their surface preparation (sandblasted and polished surface). We measured and observed the penetration of this slurry into the concrete support.
Keywords: appetency, adhesion, adherence, repair, kinetics, succion, cement, concrete.

1   Introduction

When a slurry is applied on a concrete support, bonds are developing because there is a potential attraction between the two bodies [2]. A physiological comparison let us say that there is here *partiality*. Partiality (lat. *appetentia*) is like a penchant that let the two bodies to do something they want or like to do.

The Interfacial Transition Zone in Cementitious Composites, edited by A. Katz, A. Bentur, M. Alexander and G. Arliguie. Published in 1998 by E & FN Spon, 11 New Fetter Lane, London EC4P 4EE, UK, ISBN: 0 419 24310 0

This word represents the physical, chemical and mechanical properties that will influence the macroscopical manifestation of the *adhesion*, that means *adherence*. From appentency (partiality) to adherence, there is a cause to effect relation : it is because there is a potential reactivity of the bodies that it will be possible to measure adherence by means of a test. Adhesion is the physical or physico-chemical phenomenon happening at the interface that will really produce the bonds.

Here we want to explain the influence of the kinetics of contact of the cement slurry on the concrete support on the creation of the bonds and the quality of adherence. Capillary succion is one of the most important mechanism at the base of the creation and the resistance of the interface [3].

2   Transport mechanisms at the interface

Three major transportation mechanisms are observed into porous media :

– Diffusion, related to the transfer of molecules or ions into the interstitial solution, from area of high concentration to area with lower concentration.
The phenomenon is directed by Fick's laws :

$$F = - D \frac{dc}{dx}$$

where   F is the flow rate (g/m².s)
D is the diffusion coefficient (m²/s)
c is the concentration (g/m³)
x is the distance (m).

The second Fick's law represents the variation of the concentration with the time. This transport mechanism may be present when a cement slurry is laid on a concrete support saturated with water : diffusion of ions could be observed from the interstitial solution of the slurry to the water present into porous skeleton of the concrete and inversely. The analysis of the ionic composition of the slurry is very interesting : it permits to compare its concentration value with the one of water and water into concrete.

Table 1. Ionic composition of city water (Liege, Belgium), interstitial water of concrete and centrifugated solution of a cement slurry (mg/ℓ)

Ions	Cement slurry	City water	Concrete
OH⁻	4720	-	2245
Cl⁻	1390	55.64	-
Ca	1210	145.73	945.89
Mg	< 0.05	14.24	-
Na	1500	16.49	542.56
K	7360	2.98	7897.8
pH	13.1	7.35	13.1

N.B. the values for concrete interstitial solution are coming from analysis for CPA 69 after 5 hours [5].

We may conclude that there are ions movements at the interface and not always in the same direction because the hardening of cement will modify the concentration of the different ions into water.

- Permeation, concerning the movement of liquids and gazes due to difference of pression. The mechanism is described by Darcy's law :

$$K_w = \frac{V}{t} \cdot \frac{\ell}{A} \cdot \frac{1}{\Delta h}$$

where K_w is the coefficient of water permeability (m/s)
               V is the volume of the liquid (m³)
               t is the time (s)
               ℓ is the thickness of the penetration section (m)
               A is the area of the penetration section (m²)
               Δh is the height of the water column (m).

This phenomenon supposes that there is a pressure gradient between the surface and the deepness of the concrete : this is very imprevious because the only difference could come from the force of application of the slurry or the thickness of the new layer, which is no more than 40 to 50 Pa.

- Capillary succion, related to the transport of liquids into porous solids, due to interfacial tension between liquid and solid. The phenomenon is described by a combination of Poiseuille's law and Laplace's law :

$$\ell_p^2 = \frac{r_o \cdot \gamma_{LV} \cdot \cos \theta}{2\eta} \cdot t$$

where r_o is the radius of the pore (m)
               γ_{LV} is the superficial tension of the liquid (mN/m)
               θ is the contact angle
               η is the viscosity of the liquid (mPa.s)
               t is the time (s).

This is an important and fundamental mechanism acting on the quality of the interface between the slurry and the concrete support.
It will indeed regulate the transfer of water from the slurry to the support, especially when it is dry : this will modify the hardening process of the cement and the development of adhesion at the interface. The test described in chapter 4 will try to analyse the phenomenon.

3   Description of the materials

We used prefabricated slabs of concrete made of :
- 2/8 limestone aggregate,
- 0/2 and 0/5 sand,
- CEM II B 32,5 cement and water (W/C = 0.5).

Some tests have been realised on rock itself in order to point out the influence of the cement paste. Two types of surface treatments were applied on the concrete support : the first one was a sandblasting with Corindon 1/1.4 and the second was a polishing of the surface with output of fine particles after treatment. It corresponds to a linear rugosity factor R_a respectively of 1.13 and 1.02.

Table 2. Physical and mechanical characteristics of the sandblasted and polished concrete, and limestone rock

Test	Sandblasted concrete	Polished concrete	Rock
Water absorption (% in mass)	5.49	5.38	0.27
Water absorption under vacuum (% in mass)	8.99	8.59	0.28
Total porosity (% in volume)	19.96	19.07	0.78
Specific surface (m²/g)	1.81	1.87	0.54
Porous volume (cm³/g)	0.046	0.046	0.0058
Mean radius (nm)	165	136	12

The slurry applied on the concrete support is a mix of water and cement CEM I 42,5 with W/C = 0.4 We measured the viscosity, the superficial tension and the contact angles with the supports in order to quantify and to modelize the capillary succion of this slurry into the concrete support [1].

4   Description and results of the capillar tests

The most commonly used test to analyse water transfer at the interface is the water capillar succion test [3]. It is described in a lot of standards (NBN B 14-201, DIN 52617,...) that differ often by the time when a measurement is realized. The results given hereafter (Fig. 1) present the capillar succion on concrete and rock by means of the impregnation ratio (S_t) : this is related to the evolution with time of the water capillar absorption (E_c) and the absorption under vacuum (E_t) :

$$S_t = \frac{E_c}{E_t} \times 100$$

This coefficient is defined by standard NBN B 15-201 (1976).

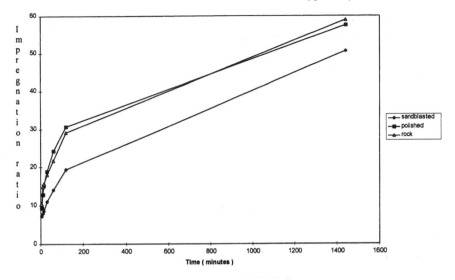

Fig. 1. Impregnation ratio for sandblasted and polished concrete surface and rock - variation in mass

The disadvantage of this test is the impossibility of realising measurement just after the contact between concrete and slurry. It is the reason why we developed a new test that register continuously the variation of mass of the sample immerged into water (Fig. 2).

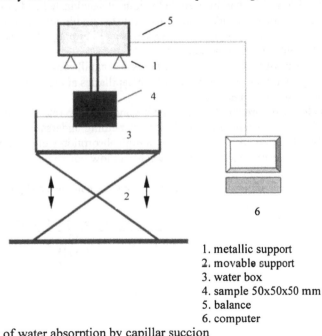

1. metallic support
2. movable support
3. water box
4. sample 50x50x50 mm
5. balance
6. computer

Fig. 2. Measurement of water absorption by capillar succion

We registered the capillary succion during more than 40 minutes and the figure 3 gives the evolution between 0 and 2400 seconds.

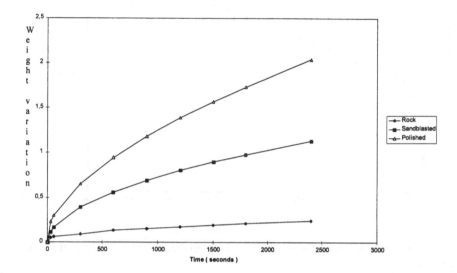

Fig. 3. Water capillar absorption between 0 and 2400 seconds on rock and sandblasted and polished surfaces – variation in mass.

We observed here a different behaviour of sandblasted and polished surfaces : the speed of absorption and the quantity of water are higher for polished one than for sandblasted one. This phenomenon is observed directly after the contact between water and concrete and may be due to :
– the shape of the entrance of the pores and capillaries;
– the granulometric distribution of the capillaries at the surface;
– the presence of a pollutant,…
It is clearly before the 20 or 30 first seconds that the rate of impregnation is the higher what is, at a theoretical point of view, something understandable : the larger capillaries are first fulfilled with water, giving a high absorption weight and after that, the thin capillaries absorb water on great height but low volume.

Table 3. Speed of the rate of impregnation of the sandblasted and polished surfaces, and rocks

Type of support	Speed of impregnation ($\% / \sqrt{t}$)
Sandblasted	0.1935
Polished	0.5518
Rock	1.1218

5   Microscopic analysis

We present here some results of S.E.M. analysis related to the interfaces between aggregates and cement paste with the slurry. These observations were realized from the analysis of the faces of failure after a pull-off test.
The distinction between the zones of contact aggregate/slurry and cement paste/slurry, particulary when the slurry was applied on a dry support, are clearly visible on photograph 2.
We note a higher roughness of the interface cement paste/slurry and a difference of structure; the aggregates appear more dark and seem to be only lightly covered by a thin layer, while the zone in contact with cement paste is covered by residual cement slurry (photograph 1).

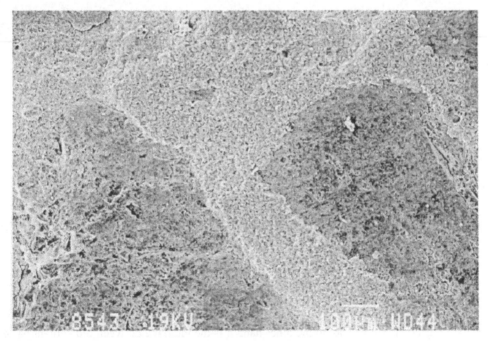

Photograph 1. Failure face of saturated polished concrete support

These observations are available as well as for polished than for sandblasted surfaces. The most effective difference between the two is coming from the air bubbles present in the surface roughness of the sandblasted support (photograph 2).
An EDX analysis attests that the thin layer present on the aggregates is in most of the cases characterized by a high concentration of $Ca(OH)_2$.

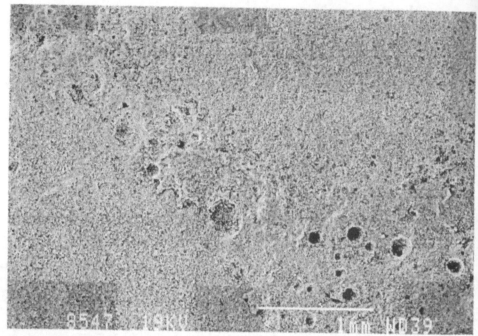

Photograph 2. Failure face of cement slurry applied on polished saturated concrete support

The effect of the saturation of the support was more clearly point out by means of fluorescent microscope analysis on thin layer. A zone of 50 – 100 μm with a higher porosity is visible at the interface, as well as between the slurry and the cement paste than between rock and cement paste : it is not clear up to now that this behaviour depends on the saturation ratio but a first analysis let us to conclude that it is more usefull for dry support.

6   Conclusions

All the observations presented hereabove let us to make a comparison with the interfacial zone between aggregates and cement paste into concrete. We may indeed consider that we are in the same conditions : on one hand it is the contact between aggregate and slurry and on the other hand it is the interface between old cement paste and slurry.

The analysis of the failure faces shows clearly that, in the first case, the rupture happened in the interfacial zone which means between the $Ca(OH)_2$ crystals as it was already described by Metha [11], Maso [12] and a lot of other authors [8] [9] [10]. In the second conditions, the rupture happens inside the slurry (new cement paste) where hardening is of course no so developed.

Repair technics, where no organic components are used, are not so different, at the point of view of the physico-chemical interactions, from concrete technics.

We shall be able to evaluate the adhesion forces available between rock and slurry and cement paste and slurry, in order to predict adherence.

Finally, adhesion and cohesion are two words for the same effect.

7 References

1. Courard, L. (1997) *Appétence et adhérence : cause et effet d'une liaison efficace.* PhD thesis, Faculty of Applied Science, University of Liege (to be published).

2. Fiebrich, M.H. (1993) Scientific aspects of adhesion phenomena in the interface mineral substrate – polymers, in *Proceedings of 2nd Bolomey Workshop : Adherence of young and old Concrete.* (ed. F. Wittmann), Unterengstringen, pp. 25-58.

3. Jutnes, H. (1995) Capillary succion of water by polymer cement mortars, in *RILEM Symposium on Properties and Test Methods for Concrete-Polymer Composites* (ed. D. Van Gemert), Leuven, pp. 29-37.

4. Larbi J., Bijen, J.M. (1991) The role of the cement paste-aggregate interfacial zone on water absorption and diffusion of ions and gases in concrete, in *The Cement paste aggregate interfacial zone in concrete* (ed. J. Bijen), Delft, pp. 76-93.

5. Longuet, P. and all (1973) La phase liquide du ciment hydraté, *revue des Matériaux de Constructions et de Travaux Publics, section Ciments/Bétons*, tome 676, Paris, pp. 35-41.

6. Tabor, D. (1981) Principles of adhesion – Bonding in cement and concrete, in *Adhesion problems in the recycling of concrete*, (ed. P. Kreijger, Nato Scientific Affairs Division), pp. 63-90.

7. Silfwerbrand, J. (1990) Improving Concrete Bond in Repaired Bridge Decks, in *Concrete International*, pp. 61-66.

8. Mindess, S. (1987) Bonding in Cementitious Composites : How important is it ? in *Bonding in Cementitious Composites*, (ed. Mindess), Pittsburgh, pp. 3-10.

9. Scrivener, K. and all (1987) Microstructural gradients in cement paste around aggregate particles, in *Bonding in Cementitious Composites*, (ed. Mindess), Pittsburgh, pp. 77-85.

10. Wang, J. (1987) Mechanism of orientation of $Ca(OH)_2$ crystals in interface layer between paste and aggregate in systems containing silica fume, in *Bonding in Cementitious Composites*, (ed. Mindess), Pittsburgh, pp. 127-132.

11. Mchta, P.K., Montciro, P.J.M. (1987) Effect of aggregate, cement and mineral admixtures on the microstructure of the transition zone, in *Bonding in Cementitious Composites*, (ed. Mindess), Pittsburgh, pp. 65-75.

12. Maso, J.C. (1980) La liaison entre les granulats et la pâte de ciment hydraté, in *VII Congrès international sur la chimie des ciments*, vol. III sous-thème 1 (ed. Septima), Paris, vol. 18, pp. 61-4.

21 RESTRAINED SHRINKAGE CRACKING IN BONDED FIBER REINFORCED SHOTCRETE

N. BANTHIA and K. CAMPBELL
Civil Engineering Department, University of British Columbia, Vancouver, Canada

Abstract
Plastic shrinkage cracking in bonded dry-mix shotcrete subjected to a severe drying environment is investigated. A novel test method is used. Shotcrete is shot straight on a fully matured, rough sub-base which provides the dimensional restraint. The assembly is then transferred to a drying chamber where early age cracking is allowed to occur in the shotcrete overlay and crack characteristics are measured. It is found that the method is very effective in estimating the potential and extent of shrinkage cracking in shotcrete, and also in assessing the effectiveness of various types of fibers and additives. Both steel and polymeric macro-fibers are investigated which are seen not only to delay the formation of cracks but also reduce crack widths and total crack areas.
Keywords: Shotcrete, interfacial bond, plastic shrinkage cracking, fiber reinforcement.

1 Introduction

Shotcrete is now recognized as one of the fastest growing materials for new construction and repairs, and has found extensive application in mining, tunneling, highway, railway and off-shore construction. Depending on the conditions at the site and the prevalent construction practice in the region, shotcrete at these projects may either be produced via the dry-process or the wet process. In the dry-process, pre-bagged materials are fed to the hopper of the shotcreting gun in the bone-dry (or sometimes partially pre-moisturized) state and conveyed to the nozzle where the mix-water is added. In the wet-process, on the other hand, fully mixed concrete including

The Interfacial Transition Zone in Cementitious Composites, edited by A. Katz, A. Bentur, M. Alexander and G. Arliguie. Published in 1998 by E & FN Spon, 11 New Fetter Lane, London EC4P 4EE, UK, ISBN: 0 419 24310 0

all the required mix water is added to the hopper of the gun and pumped to the nozzle where depending on the type of the machine additional air may be added for an increased material velocity on exit.

One of the major problems with the dry-process shotcrete is the high aggregate and fiber rebound; losing up to 50% of each of these components through rebound is not uncommon. Further, large aggregate particles have a tendency to rebound as much as four times the small aggregate particles. This results in an increase in the cement content in the in-place shotcrete sometimes by as much as a factor of two. The loss of fiber through rebound, on the other hand, may reduce the *in-situ* fiber volume fraction by as much as half. Both of these have obvious implications in terms of cost and performance. To make matters worse, in many instances, site conditions do not also permit an adequate curing of in-place shotcrete further affecting the performance.

With a very high cement content and inadequate curing, early age shrinkage cracking in dry-mix shotcrete is one of the major concerns. Fiber reinforcement may be expected to control shrinkage induced cracking, but the excessive loss of fiber through rebound diminishes their *in-situ* effectiveness. Premature cracking in shotcrete is seen all too often, and this is especially true in cases where a strong interfacial bond develops between the substrate and the shotcrete overlay. A strong interfacial bond imposes a severe dimensional restraint on shotcrete and cracking becomes the only way of dissipating the strains developed through shrinkage. Additional factors that may be expected to control the possibility of cracking in shotcrete are gradation, type and properties of aggregate, properties of mineral and chemical admixture present, efficiency of the fiber employed, severity of the restraint and the severity of the environment. On the process side, rebound of both fibers and aggregate is known to be affected by the nozzle type, nozzle distance from the surface, nozzle angle, air pressure, experience of the nozzleman, shooting consistency, etc. and these, in turn will also all control the extent of cracking during shrinkage.

2 Experimental

Unfortunately, there are no standardized procedures for assessing the potential for cracking in restrained concrete or shotcrete in a drying environment. Imposing a valid dimensional restraint is often the problem. A number of diverse techniques, however, have been developed including the shrinkage ring (1) and various other linear specimen arrangements (2,3). In principle, however, most of these procedures apply only to cast concrete and are not suitable for shotcrete. In the case of shotcrete, extreme heterogeneity exists at the edges due to high proportion of trapped rebound at these locations and hence both narrow elongated specimens or ring-type specimens do not represent reality. In shotcrete research, it is highly uncommon to shoot beams directly; only panels are shot such that the edges can be later trimmed off to retrieve linear beam specimen through sawing.

The other difficulty is the nature of the physical restraint imposed during a test. In the existing techniques, generally an "end" or an "edge" restraint is imposed, but in most applications, shotcrete is not restrained in this way.   Shotcrete generally develops a strong bond with the stabilized soil or rock substrate, and hence the restraint imposed is along the entire interfacial region developed between the bottom surface of the shotcrete overlay and the substrate. This can be further aided by the local restraint imposed by pins, anchor bolts and anchor plates.

In an earlier study, a novel test method was devised to study the influence of drying on the shrinkage cracking in cast concrete (4). The same was adopted in this study for shotcrete.   Dry-process shotcrete was produced using a rotary barrel equipment (ALIVA 246) with a 3.6 liter 8 pocket drum (Fig. 1). This equipment is instrumented with a spring loaded in-line airflow meter (OMEGA FL8950). A 20 m long 50 mm internal diameter hose was used to convey the material to the nozzle, and the mix water was added 2m before the nozzle at a high pressure between 1 MPa and 5.2 MPa controlled by a manometer operated at the nozzle. The gun was operated at the maximum rotor speed for all shoots.

Fig. 1. Dry-mix shooting in progress on a rough sub-base.

Shotcrete was shot directly on a prepared sub-base (300 mm × 900 mm) cured to maturity for at least 90 days and left intentionally with a rough 'exposed' aggregate finish (Fig. 1). Once applied, the surface of shotcrete was finished by a trowel, and the whole assembly was transferred to a drying chamber at approximately 40°C and 10% relative humidity (Fig. 2). In the chamber, as and when needed, three fans blew hot air on top of the composite specimen. The shotcrete overlay was demoulded 1h after casting and crack observations were begun with a 100× optical microscope. More specifically, crack widths and crack lengths were measured to an accuracy, respectively of ±0.025 mm and ±5 mm. Throughout a test, a rate of evaporation of about 1.5 kg/m^2/h was maintained in the chamber. This was measured by placing a bowl of pure water in the chamber and by monitoring its mass as a function of time. Clearly, this is only an approximate measure, because the actual rate of evaporation from the shotcrete surface would depend upon the rate at which water becomes available.

Fig. 2. The environmental chamber.

As expected, the roughness of the sub-base would also greatly influence the cracking in the shotcrete overlay. For a consistent roughness, crushed aggregates 19 mm in average size were carefully left exposed to half their size with an aggregate to aggregate distance of 25 ± 10 mm.

Results with two types of steel fibers (S_1: crimped geometry, crescent shape 32 mm long; S_2: hooked ends, 30 mm long and 0.5 mm in diameter) and one type of polyolefin fiber (P, straight undeformed, 25 mm long and 0.38 mm in diameter) are presented here at various fiber volume fractions.

3 Results and Discussion

Only a brief description of the test results appears here. Detailed description and the development of a fracture-based model may be found elsewhere (5).

The observed crack patterns for plain shotcrete are shown in Fig. 3. The changes in the crack pattern with an addition of 15 and 60 kg of the Fiber S_2 are shown in Figs. 4 and 5. Notice that substantial cracking occurred in shotcrete in the absence of fibers, but addition of fibers was helpful in substantially reducing the shrinkage cracking. Also, the extent of cracking reduced significantly with an increase in the fiber volume fraction such that at 60 kg/m^3 fiber dosage, the cracking was completely eliminated.

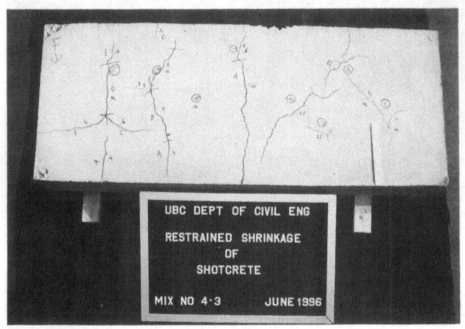

Fig. 3. Observed crack pattern in plain shotcrete

In Fig. 6, total crack areas are plotted for the fibers tested as a function of the fiber volume fraction. The increased effectiveness of the Fiber S_2 over both the crimped Fiber S_1 and the polymeric Fiber P in controlling shrinkage cracking is noticeable. It

Fig. 4. Observed crack pattern in shotcrete with S_2 (hooked-ends) fiber at 15 kg/m^3

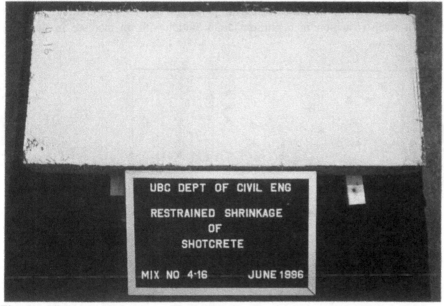

Fig. 5. Observed crack pattern in shotcrete with S_2 (hooked-ends) fiber at 60 kg/m^3. Notice a complete elimination of cracking.

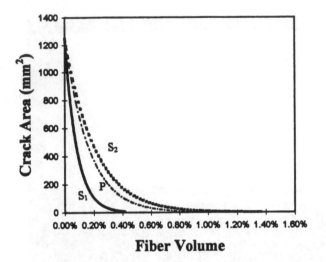

Fig. 6. Total crack areas plotted as a function of fiber volume fraction for various fibers.

is worthwhile to note here that polypropylene has a specific gravity almost 8.7 times lower than that of steel which means that on a unit mass reinforcement basis, polypropylene fiber may be just as effective as steel fiber in controlling cracking.

The maximum observed crack widths are given in Fig. 7. The trend indicates that there is a steep reduction in maximum crack width with an increase in the fiber

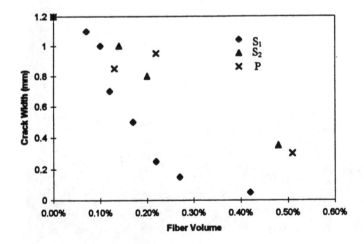

Fig. 7. Maximum crack widths plotted as a function of fiber volume fraction for various fibers.

volume fraction. Shotcrete with fiber reinforcement hence can be expected to posses significantly enhanced durability performance and act as an effective barrier to aggressive agents capable of corroding the embedded steel anchor bolts, etc.

Finally, the data clearly indicates that the novel test procedure adopted here for measuring and studying shrinkage cracking in shotcrete has significant merit and may serve as an effective tool for further studying the problem. Nevertheless, a number of issues remain unresolved. These include the possible influence of specimen size, specimen geometry, roughness of the sub-base, time of demoulding, type of surface finish and the specific conditions of drying and air velocity in the chamber.

4 Conclusions

1. A novel technique of investigating plastic shrinkage cracking in dry-mix shotcrete is presented. Results indicate that the technique has sufficient merit and is very effective in generating cracking in bonded shotcrete in a consistent manner.

2. When evaluated using this technique, fibers are found to be highly effective in reducing crack widths and overall crack areas during shrinkage. The geometry of the steel fiber appears to have an influence on its crack arrest capabilities. Polymeric fibers may be just as effective in controlling plastic shrinkage cracking as steel fibers.

5 References

1. Kovler, K., Sikuler and Bentur, A., Restrained Shrinkage Tests of Fiber Reinforced Concrete Ring Specimens: Effect of Core Thermal Expansion, Materials and Structures, 1993, 62, pp. 213-237.

2. Bloom, R. and Bentur, A., Free and Restrained Shrinkage of Normal and High Strength Concrete, ACI Materials Journal, 92(2), 1995.

3. Banthia, N. and Azzabi, M. and Pigeon, M., Restrained Shrinkage Cracking in Fiber Reinforced Cementitious Composites, Materials and Structures RILEM (Paris), 26 (161), 1993, pp. 405-413.

4. Banthia, N., Yan, C. and Mindess, S., Restrained Shrinkage Cracking in Fiber Reinforced Concrete: A Novel Test Technique, Cement and Concrete Research, 26(1), 1996, pp. 9-14.

5. Campbell, K., Shrinkage Cracking in Dry-Mix Shotcrete, M.A.Sc. Thesis, The University of British Columbia, 1997, *in preparation.*

22 FEATURES OF THE INTERFACIAL TRANSITION ZONE AND ITS ROLE IN SECONDARY MINERALIZATION

D. BONEN
USG Corporation, Research Center, Libertyville, Illinois, USA

Abstract
The notion that a definite interfacial transition zone (ITZ) some 30 to 50 µm thick is formed between aggregate and paste is challenged. Microstructural studies show that the ITZ is neither continuous nor uniform. Furthermore, the appearance of the paste near the aggregate as compared to the bulk of the paste discloses marginal microstructural differences. Nonetheless, it appears that the near interfacial zone (< 10-15 µm) bears some attributes that distinguishes it from the surrounding by the occurrence of a higher amount of calcium hydroxide and larger pores. The association of calcium hydroxide, large pores, and nucleation sites makes the near interfacial zone vulnerable to any kind of chemical attack. Despite the non-uniformity, the role of the near interfacial zone is best manifested by secondary mineralization. There is a consistent pattern indicating that the near interfacial zone offers little resistance to solution transport and precipitation. Under corrosive conditions, calcium hydroxide is the first to react out to form deposits along aggregate boundaries. Common thickening of secondary mineralization exerts stresses that facilitate degradation.
Keywords: cement microstructure, deposition site, interfacial transition zone, secondary mineralization

1 Introduction

To date a large body of literature deals with the various effects of the interfacial transition zone (ITZ) developed between the paste and aggregate. The notion that concrete is a three-phase composite consisting of cement paste, aggregate, and ITZ has become so popular that attempts to assign measurable properties to it have been made[1,2,3]. Mindess[1] has reviewed the literature and pointed out to the inconsistencies in the ITZ properties. At large this has been attributed to differences in specimen geometries and test procedures. In turn, the difficulties in determining the ITZ properties might also be related to assumption that the ITZ of a given concrete has a specific microstructure that lends certain properties. The question arises to what extent

The Interfacial Transition Zone in Cementitious Composites, edited by A. Katz, A. Bentur, M. Alexander and G. Arliguie. Published in 1998 by E & FN Spon, 11 New Fetter Lane, London EC4P 4EE, UK, ISBN: 0 419 24310 0

such an assumption is justifiable? Otherwise, what are the attributes that characterizes this zone?

This paper reviews the microstructure of the interfacial zone as observed under SEM with an attempt to address the uniformity of this zone and to point out at some consistent patterns. It happens that the attributes of the ITZ are not necessarily related to a uniform microstructure, rather, the most prominent feature of this zone is related to secondary mineralization.

2 Material and Experimental Procedure

Examinations were carried out on a number of specimens obtained from different suites of field concrete that have been in service for several years. In addition, other lab mortar/concrete specimens are also included. The microstructural analysis was carried out in SEM, mainly in a backscatter electron mode on flat polished surfaces. Phase were identified either according to their specific gray level or where the former method could be inconclusive, EDX spot analyses were used for an unambiguous characterization.

3 Features of the near interfacial transition zone

According to the "classical view" the ITZ extends some 30-50 μm away from the aggregate surface into the paste. This zone is distinguished from the bulk of the paste by greater porosity, greater calcium hydroxide and ettringite deposits, and less amount of unhydrated cement grains[4,5,6]. Further investigations, primarily in a backscatter mode indicated that the calcium hydroxide content in the ITZ is marginally greater than that of the bulk of the paste[7].

The notion that the ITZ occupies a fairly wide zone has first been criticized by Diamond[8] who pointed out that the average distance between adjacent aggregate in ordinary concrete is less than 100 μm. Thus, if the ITZ extends some 30-50 μm from the aggregate much of the bulk of the paste should be considered as ITZ. Furthermore, applying image analysis technique and measuring consecutive 10 μm wide strips around various aggregate grains, Diamond[9] noted that out of the three variables quantified: porosity, calcium hydroxide content, and unhydrated cement grains, only the latter one varies consistently. Marginal differences in porosity were also encountered. However, it has been found that variations between adjacent aggregate particles are greater than the variations across the ITZ surrounding the individual aggregates.

Previous microstructural investigation on this subject has led the current author to suggest that the visible variations that distinguishes the ITZ from the bulk of the paste occur in a narrow zone that has been referred to as the near interfacial zone[10]. This zone is distinguishable by a greater amount of calcium hydroxide deposition and large pores. No specific enrichment of ettringite or other calcium-sulfate aluminate phases has been recorded. Similarly, no specific width has been assigned to the near interfacial zone as it varies from a few microns to about 10-15 μm. Further away differences are questionable.

In agreement with Diamond's observations, large microstructural variations are recorded between adjacent aggregates. In addition, neither the calcium hydroxide content

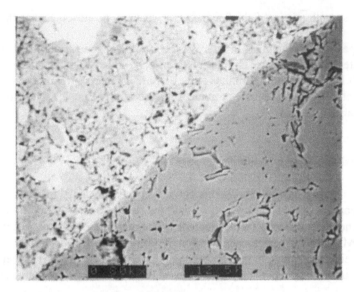

Figure 1. Backscatter electron image (BEI) illustrating typical microstructure. Along the aggregate boundary a discontinuous calcium hydroxide layer has been formed. Large pores are associated with this layer or next to it.

and its texture nor the porosity next to the aggregate surface is uniform, and changes occur on a microscopical scale along the aggregate boundaries.

The variations along the aggregates' boundaries and the different modes of calcium hydroxide deposition have been detailed in ref. 10. Preferred sites of calcium hydroxide depositions are around aggregate grains. Generally speaking, the thickness of such deposits varies from 1 to 5 μm, but frequently deposits 10 μm thick are encountered. On many other occasions, however, calcium hydroxide thins out and is practically missing. Figure 1 provide an example of dolomite aggregate on the lower right and the paste on the upper left. Along the aggregate boundary a coating of a calcium hydroxide layer ranges from 1 up to 6 μm thick has been formed. It is evident, however, that this coating is neither uniform nor continuous. Another conspicuous element is the large visible porosity that is associated with or appears next to the calcium hydroxide deposit. Apart from these two features, the appearance of paste in the near interfacial zone is very similar to that of the bulk of the paste some 50 μm away from the dolomite particle.

Attention is called to the presence of fully or partially hydrated phenograins[11] next to the dolomite particle. Where an open space is available, despite the "wall effect", sizable clinker grains can be packed against the aggregate surface. However, in confined spaces where the distance between adjacent aggregates is small, the paste becomes more porous. Figure 2 shows the microstructure formed in a space confined between dolomite particle on the left and granite one on the right situated about 80 μm apart. Calcium hydroxide lines both aggregates. It forms semi-continuous layer along the dolomite 1 to 10 μm thick. However, no definite layer is observed along the granite on the right.

A plot of the elemental variations of Ca (bottom), Si (middle), and Mg (top) measured along a line traverse is given in on the right of Fig. 2. The trace of the line traverse is depicted on the backscatter electron image on the left at about 1/4 from the top. The line

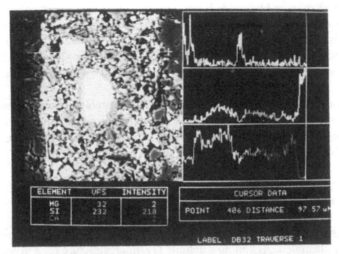

Figure 2. BEI micrograph of confined space and the elemental variations of Ca(bottom), Si(middle), and Mg(top) measured along a line traverse from the dolomite particle on the left to the granite on the right. Note the difference in the mode of calcium hydroxide deposition along the surface of the two particles and the porous area formed next to the granite particle.

traverse was carried out by conducting 512 individual X-ray spot analyses measured at intervals of 0.19 μm apart from the dolomite particle on the left to the granite particle on the right. Interpretation of the elemental patterns suggests that the phases present in the paste from left to right comprise a layer 8.2 μm thick of calcium hydroxide next to the dolomite particle, followed by 26 μm of partially hydrated C_2S phenograin, 37.4 μm groundmass, and about 13μm of porous groundmass that is increasingly enriched with calcium as the granite surface is approached. Examination at higher magnification reveals that the calcium enrichment is due to an intimate association of disseminated minute calcium hydroxide grains and groundmass.

Despite the small step size of the line traverse, the transition from both aggregate particles to the paste is anything but sharp and occurs over a few microns. In part, the gradual change is certainly related to the volume sample of the X-ray generated that is much larger than the step size of the traverse. It may also indicate, however, that the transition zone sensu stricto is quite porous. To this end, based on Auger electron spectroscopy patterns Roy and Jiang[12] have suggested that a thin layer less than 6-7 μm is formed between the aggregate and paste pointing at a gradual transition from the aggregate to the paste.

It should be noted that only part of the aggregate particles are accompanied with calcium hydroxide. Analysis of 100 aggregate particles composed of different rock types, e.g., quartz, limestone, dolomite, granite, and magmatic rock, shows that about 25% does not have calcium hydroxide deposits. Out of the remaining 75%, about half is accompanied with non-uniform but semi-continuous layer at least along one side of the circumference, whereas the other half has a non-continuous calcium hydroxide deposits. In all, the probability to find no or a non-continuous layer of calcium hydroxide in the near interfacial zone appears to be greater than encountering a semi-continuous layer.

Precipitation of calcium hydroxide is not limited to aggregate's surface only. Rather, calcium hydroxide lines air voids and is known to be associated with fibers, especially hydrophilic one such as fiber glass[13]. Of interest is the occurrence of calcium hydroxide deposits about 5μm thick along hydrophobic polypropylene fibers indicating that nucleation sites could be any inclusion whether it is an aggregate particle, fibers, or an air void.

As a final note, the ongoing discussion emphasizes the ambiguity associated with the definition of the ITZ. Different attributes have different ranges, e.g., the actual aggregate-paste transition has a short range, followed by calcium hydroxide deposition, and large pore distribution.

4 Porosity and secondary mineralization in the near interfacial zone

In the previous section the non-uniformity of the ITZ that takes place on a microscopic scale has been illustrated. Assigning a definite ITZ microstructure for a given concrete appears therefore inadequate, especially as this microstructure can be locally affected by bleeding water or uneven distribution of mineral admixture such as silica fume that densifies this zone[14]. Nonetheless, regardless of the inhomogeneity two features are likely to be encountered there: calcium hydroxide deposition and large pores.

Indirect measurements of porosity conducted by Winslow et al[15] have shown that incorporation of aggregate results in additional porosity that occurs in pore sizes larger than the plain paste's threshold diameter as measured by mercury intrusion technique. In turn, the increase of the pore's threshold increases the permeability[16]. A further increase of permeability occurs if the calcium hydroxide leaches out from this zone. According to a digital-based simulation model, Bentz and Garboczi[17] suggested that the diffusivity might be increased by an order of magnitude.

The porosity associated with the near interfacial zone is probably related to the hydrophilic nature of the aggregate and the inefficient packing of the cement particles because of the wall effect[18]. Upon mixing, a film of water is formed on the aggregate surface. As the water is taken and the volume of the groundmass generated is insufficient to fill up the previous water-filling spaces pores are left behind. This residual porosity is also related to an inferior ion transport that is restricted to two directions, thus interfering and hindering formation of groundmass that could densify this zone. In confined spaces (such as observed in Fig. 2) this phenomenon is further augmented as the ion transport is limited to one direction only.

In view of its porous nature, the near interfacial zone plays a major role in secondary mineralization. Regardless of the environmental conditions, secondary mineralization is perhaps the most prominent feature of the ITZ as it follows a consistent pattern.

Portland cement is prone to various types of corrosion. Aggressive solution attack on cement-based materials, results in a considerable amount of dissolution and precipitation[19]. In all of these reactions, Ca^{2+} is an essential ingredient, as it is needed for the formation of secondary minerals such as ettringite, gypsum and various forms of carbonates. The vulnerability of the near interfacial zone to secondary mineralization and degradation is attributed to the association of calcium hydroxide deposits, large pores, and availability of nucleation sites. As any leaching or ingress process involves substance transport, the additional porosity in large pore sizes makes the near interfacial

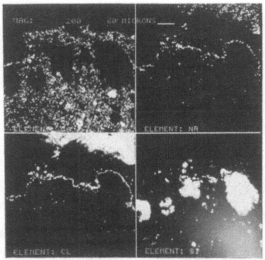

Figure 3: X-ray maps of Ca (upper left), Na (upper right), Cl (lower left), and Si (lower right). A distinct continuous layer of NaCl accompanies the aggregate boundaries. The white bar displayed at the top of the Na map corresponds to 60 μm.

zone more likely to be affected. The following examples demonstrate this point; first let's review deposition sites of innocuous secondary mineralization.

Figure 3 is a X-ray map of Salado Mass Concrete placed in thick salt layers[20]. The image shows the elemental distribution of calcium (upper left), sodium (upper right), chlorine (lower left), and silicon (lower right). Both sodium and chlorine maps delineate the distribution of halite (NaCl). The calcium maps shows the occurrence of the hardened cement paste and that of silicon the location of the aggregate. The position of the salt host rock is seen on the upper right of the chlorine and sodium maps. A distinct

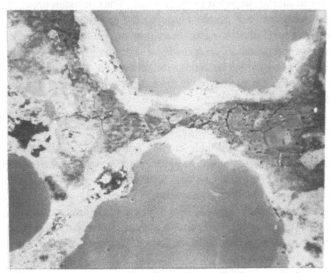

Figure 4: BEl micrograph showing deposition of rebar corrosion products along aggregate boundaries.

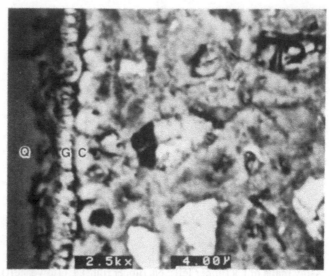

Figure 5: BEI micrograph showing a double-layer coating along quartz (Q) grain. The coating comprises of gypsum (G) and calcium hydroxide (C). Gypsum didn't deposit within the bulk of the paste.

continuous halite layer a few microns thick has been deposited in along the boundaries of aggregates of various types. It is evident that the intensity of the Na and Cl signals along the aggregates is greater than the intensity found in the paste.

Examination at high magnification reveals that many of the individual halite crystals have a rhythmic zoning suggesting a growth over time. It means that this zone has been subjected to a prolonged period of ion transport.

In contrast to the previous example in which precipitation is related to an external solution ingress, Fig. 4 shows deposition sites associated with an internal migration took place in reinforced mortar subjected to NaCl attack. The corrosion products of a near-by rebar have been mobilized and much of it deposited around the quartz aggregate rather than the bulk of the paste illustrating again that the near interfacial zone is a favourable site for secondary mineralization. In accordance with the previous discussion, confined spaces offer higher porosity that becomes available for secondary mineralization. A strip about 30 μm wide of reaction products is formed between the aggregate particles at the lower part of Fig. 4.

The susceptibility of the near interfacial zone is further augmented when the aggressive solution reacts with calcium hydroxide. Figure 5 provides a rare image of a partial replacement of the calcium hydroxide coating due to magnesium sulfate attack. The aggressive solutions penetrated along the aggregate surface and reacted with the calcium hydroxide to produce a continuous gypsum layer at the interface. As not all the calcium hydroxide reacted out, a double-layer coating is formed along the aggregate boundary composed of an inner gypsum layer followed by the remaining calcium hydroxide one. Since the bulk of the paste does not contain gypsum, it supports the view that the near interfacial zone is more permeable than the bulk of the paste and the calcium hydroxide there is more likely to react out.

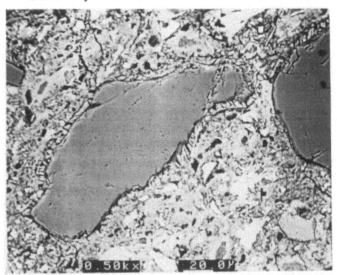

Figure 6: BEI micrograph of a string of massive-type ettringite coats aggregates.

Crystallization does not happen randomly but is triggered by nucleation sites such as aggregate surface and a previous occurrence of deposition. A common feature concerns with a thickening of previous deposits. Once mineralization takes place, often the layer generated around the aggregate thickens to form deposits up to 100 μm wide. Evidently, such thick deposits generate stresses and result in radial and concentric cracking[21,22].

At an advanced stage of distress related to delayed ettringite formation characteristic strings of massive ettringite are formed. It has been shown that this massive ettringite has an almost stoichiometric composition whereas small disseminated ettringite in the paste is highly non-stoichiometric[23]. Such strings of ettringite are associated with aggregate boundaries and quite often string run across the paste from one aggregate particle to the next one engulfing each of the particles with a massive deposits of tubular ettringite (Fig. 6). This type of ettringite is formed through extensive dissolution and mobilization of calcium, sulfur, and aluminum. Regardless of the mechanism of degradation, it highlights the role of the aggregate interface in degradation processes due to high ion transport and extensive secondary mineralization.

5 Concluding Remarks

The interfacial transition zone is neither uniform on a macro scale nor on a micro one. This non-uniformity makes it meaningless to assign to it a definite microstructure. Likewise, the assumption that the ITZ can be regarded as a single property shell formed around aggregate is not supported by microstructural examination. Nonetheless, two attributes characterize the near interfacial zone extending some 10-15 μm away from the aggregate surface: occurrence of calcium hydroxide and large pores. The porous nature of the near interfacial zone coupled with availability of nucleation sites makes it vulnerable to solution ingress and a preferred site for secondary mineralization. As a result, the role of the interface in degradation processes becomes apparent.

6 References

1. Mindess, S. (1996) Mechanical Properties of the Interfacial Zone: A Review. *ACI Special Publications,* Vol. 156, pp.1-10.

2. Garboczi, E.J. and Bentz, D.P. (1996) The effect of the interfacial transition zone on concrete properties. In *Proc. Materials Engineering Confer.* ASCE, Vol 2, pp. 1228-1237.

3. Lutz, M.P., Monteiro, P.J.M. and Zimmerman, R.W (1997) Inhomogeneous interfacial transition zone model for the bulk modulus of mortar. *Cem. Concr. Res.* Vol. 27, No. 7. pp 1113-1122.

4. Farran, J. (1956) Contribution mineralogique a l'Etude de l'Adherence entre les constituants hydrates des ciment et les materiaux enrobes. *Rev. Mater. Construct.,* No. 490-491,492.

5. P. J .M. Monteiro, P.J.M. and P. K. Mehta, P.K. (1985) Ettringite formation on the aggregate-cement paste interface. *Cem. Concr. Res.,* Vol 15 No.2 pp. 378-380.

6. Scrivener K.L. and Gartner, E.M. (1988) Microstructural gradients in concrete paste around aggregate particles. in *Bonding in Cementitious Composites,* (ed. S. Mindess and S. P. Shah) Proc. Mat. Res. Soc. Vol. 114, Pittsburgh, PA, pp.77-86.

7. Scrivener, K.L. (1989) The microstructure of concrete. in *Material Science of Concrete.* (ed by J. Skalny), Am. Ceram. Soc. Westerwille, OH, pp.127-161.

8. Diamond, S. (1986) The Microstructure of cement paste in concrete. *Proc. 8th Inter. Cong. Chemistry of Cement,* Rio de Janeiro, Vol. I pp. 122-147.

9. Diamond, S. (1977) see this publication.

10. Bonen, D. (1994) Calcium hydroxide deposition in the near interfacial zone in plain concrete. *J. Am. Ceram. Soc.* Vol. 77 No. 1, pp. 193-196.

11. Diamond, S. and Bonen, D. (1995) A re-evaluation of the microstructure of hardened cement paste based on backscatter SEM results. in *Microstructure of Cement-Based Systems/Bonding and Interfaces in Cementitious Materials,* (eds. S. Diamond, F.P. Glasser, L.W. Roberts, J.P. Skalny, and L.D. Wakeley), Proc. Mat. Res. Soc. Vol. 370, Pittsburgh, PA, pp. 13-22.

12. Roy, D.M. and Jiang, W. (1995) Influences of interfacial properties on high-performance concrete composites, in *Microstructure of Cement-Based Systems/ Bonding and Interfaces in Cementitious Materials,* (eds. S. Diamond, F.P. Glasser, L.W. Roberts, J.P. Skalny, and L.D. Wakeley), Proc. Mat. Res. Soc. Vol. 370, Pittsburgh, PA, pp. 309-318.

13. Bentur, A. and Mindess, S. (1990) Fiber reinforced cementitious composites, Elsevier, London.

14. Goldman, A. and Bentur, A. (1989) Bond effects in high strength silica fume concretes. *ACI Materials J.* Vol. 86, No. 5, Sept.-Oct., pp. 440-447.

15. Winslow, D.N., Cohen, M.D., Bentz, D.P., Synder, K.A. and Garboczi, E.J. (1994) Percolation and pore structure in mortar and concrete. *Cem. Concr. Res.* Vol. 24, No.1, pp. 25-37.

16. Halamickova, P., Detwiler, R.J., Bentz, D.P. and Garboczi, E.J. (1995) Water permeability and chloride ion diffusion in portland cement mortars:relations to sand content and critical pore diameter. *Cem. Concr. Res.* Vol. 25 No. 4, pp-790-802.

17. Bentz, D.P. and Garboczi, E.J. (1992) Modeling the leaching of calcium hydroxide from cement paste:effects on pore space percolation and diffusivity, *Materials and Structures*, 25 pp.523-533.

18. Bentz, D.P., Garboczi, E.J., and Stutzman, P.E. (1993) Computer modelling of the interfacial zone in concrete. in *Interfaces in Cementitious Composites*, (ed. J.C. Maso) RILEM Proc. 18, E&FN Spon, London, pp. 107-116.

19. I. Biczok, Concrete Corrosion and Concrete Protection; p. 543, Akademiai Kiado, Budapest, 1964.

20. Bonen, D, (1977) The microstructure of concrete subjected to high-magnesium and high-magnesium sulphate brine attack," 8 pp in *The 10th International Congress on the Chemistry of Cement*, Gothenburg, Sweden,Vol 4iv022 pp 8.

21. Bonen, D. (1993) A microstructural study of the effects produced by magnesium sulfate attack on plain and silica fume-bearing portland cement mortars. *Cem. Concr. Res.* Vol. 23 No. 3, pp. 541-53.

22. Bonen, D. and Sarkar, S.L. (1993) Replacement of portlandite by gypsum in the interfacial zone and cracking related to crystallization pressure, in *Cement-Based Materials: Present, Future and Environment Implications*, (eds. M. Moukwa, S.L. Sarkar, K. Luke, and M.W. Grutzeck) Ceramic Transactions Vol. 37, Am. Ceram. Soc. Westerville, Ohio, pp. 49-59.

23. Bonen, D. and Diamond, S. (1977) Characteristics of delayed ettringite deposits in ASR-affected steam cured concretes. in *Mechanism of Chemical Degradation of Cement-Based Systems*, (eds. K.L. Scrivener and J.F. Young), E & FN Spon, London, pp. 297-304.

23 ASSESSMENT OF CEMENTITIOUS COMPOSITES BY IN-SITU SEM BENDING TEST

P.J.M. BARTOS and P. TRTIK

ABSTRACT

This paper presents the first results of a new test method - SEM in-situ 3-point bending test. Newly designed test rig which fitted previously used eucentric SEM in-situ pull-out chamber allowed the authors to observe GRC microfracture mechanisms on specially designed specimens subjected to three-point bending test. In combination with microstrength tests, this method offers new means of investigation into fracture mechanics of cementitious composites reinforced by bundled reinforcement.

KEYWORDS

SEM in-situ three-point bending test; bundled reinforcement; GRC; crack propagation; microstrength test;

INTRODUCTION

As shown in the previous studies carried out at University of Paisley e.g. (Bartos & Zhu; 1997) a new mode of failure of bundled reinforcement in cementitious composites, specifically in GRC, had been discovered. The "telescopic" mode of failure of bundled reinforcement proved to be highly desirable for practical GRC and opened new avenues for development of composites reinforced with bundles of fibres. Within the framework of this investigation, a new test method, an in-situ 3-point bending test, was developed, which provided an additional tool to currently used microstrength test for assessment of micromechanisms of fracture in cementitious composites.

In-situ SEM tests have been used for obtaining the information about micromechanisms of fracture in cement based composites since early 1980's. Major pioneering work was carried out by Mindess and Diamond (1980) and Lovell & Diamond & Mindess (1983), who introduced both wedge opening loading and compression test methods. It is also necessary to mention the work of Tait and Garrett (1985), who developed and utilised double torsion method. The micromechanisms of bundled reinforcement in cementitious composites observed via the in-situ SEM test methods were described by Bentur & Diamond (1983). At the University of Paisley the in situ tests SEM have been widely used by Swift (1987), Duris (1993) and Zhu (1996), in all cases in the form of either tensile or fibre pull-out tests. The advantages and disadvantages of the test methods mentioned above and their comparisons with the test method presented are discussed later in this paper.

EXPERIMENTAL

Test arrangement

The tests were carried out inside the special chamber previously developed at the University of Paisley. This chamber replaces the ordinary specimen chamber on the CamScan SEM. The chamber is fitted with load cell of maximum load capacity of 500 N. The test stage allows the eucentric movements of grips with a controlled velocity from 2μm/sec to 20μm/sec. A newly designed rig enabled the authors to perform 3-point bending test. A new data capture system was fitted to the chamber as well thus providing output in a digital form. During the testing, the value

The Interfacial Transition Zone in Cementitious Composites, edited by A. Katz, A. Bentur, M. Alexander and G. Arliguie. Published in 1998 by E & FN Spon, 11 New Fetter Lane, London EC4P 4EE, UK, ISBN: 0 419 24310 0

of the instantaneous load was captured in interval of one second. Typical set-up of the testing chamber prepared for the bending test is shown on Figs 1 and 2.

Figure 1 - SEM in-situ three point bending test stage

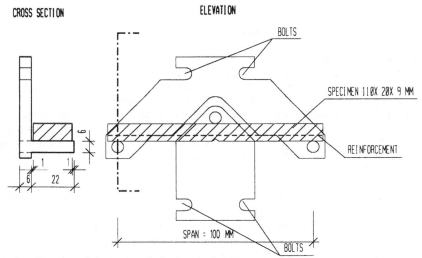

Figure 2 - Drawing of the testing rig for 3-point bending test

Arrangement of microindentation apparatus which has been used for microstrength tests is shown on Fig.3. All the details concerning the basic operating principle of this apparatus can be found in the reference (Bartos, 1997).

Since the microindentation tests have been performed within this project as well, the test arrangement and the basic operating principles of the microindentation apparatus on Fig. 3.

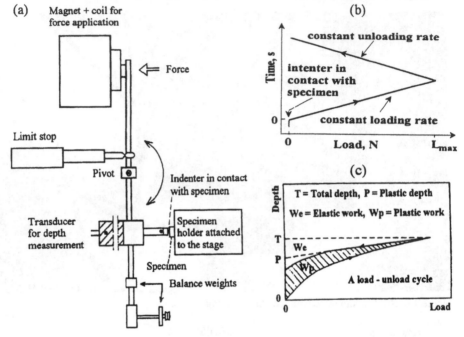

Figure 3 - Test arrangement of the micro-indentation apparatus

Specimen production

Specimens which have been used within the project were prisms 110x20x9 mm in size containing one layer of parallel glass fibre strands. The fibre reinforcement was positioned 3 mm above the bottom edge of the specimens. Twelve strands were spaced equally across the 20 mm width of the prism in such a manner that the inner strands (10 in number) were fully embedded in the matrix whilst two outer strands have been placed on the side of the prism thus being partially embedded only. This particular arrangement of strands was chosen to suit the test method. The partially-embedded strands have been visible on the surface of the specimen which allowed the authors to observe the bridging effect of fibre bundle during the crack propagation. Alkali resistant NEG-AR fibres were used in this project. The overall fibre content achieved was about 0.3% by volume and the fibres represented a case of a continuous reinforcement. A matrix of a constant composition was used for all the specimens tested. In order to avoid the necessity to compact the matrix by vibration a cement slurry which had been developed for slurry infiltrated fibre concrete (Marrs, 1996) was used in this project. The mix proportions were: 1000 g of Blue Circle Portland Cement 42.5R, 1000 g of fine aggregate (passing 600 microns sieve), 550 ml of tap water, 15 ml of superplasticizer Sikament, 5 g of underwater concrete admixture Conplast. The mould for the specimen preparation which was specially designed for previous research at the University of Paisley has been utilised for this project. In order to ensure the accurate positioning of the strands within the specimen a detachable frame with notches and a double sided tape were attached to the bottom plate of the mould. After the bundles of fibres had been assembled into the mould the matrix was carefully poured in. The specimens were covered by a polyethylene sheet and they were demoulded after 48 hours. Fibres protruding from the specimens were cut off. The specimens were marked for identification and placed in 20°C water for the initial curing. After five days all specimens were removed from the water. The specimens selected for the accelerated ageing were transferred into a hot water bath (60°C) and they were kept there for 21 days. The aged specimens were notched. The aim was to produce identical notches in each specimen. Difficulties occurred due to the requirement of the highest possible sharpness of the tip of the notch. First, the base of the notch was carefully cut by a fine hand saw, then the notch was "engraved" by means of a sharp steel tool. The tip of the notch was "finished" by means of a map pin. The result was a V-shaped notch about 1.5 millimetres deep. For a successful SEM observation, the specimens had to be dried in the oven for three days at the temperature of 75°C. The specimen preparation procedures were completed by a gold coating which ensured a much better image in the scanning electron microscope.

Test pieces for microstrength tests were produced by cutting the specimens which underwent the 3-point bending test perpendicularly to their longitudinal axis. The specimens were 5mm thick. A very fine saw was used for this purpose and propan-2-ol liquid was used for lubricating the cut surface during this procedure. Thus prepared, the test-piece was then mounted on the holder by means of superglue.

RESULTS
The test results of bending tests confirmed the authors' expectations concerning the behaviour of OPC-matrix GRC in bending. Typical load-deflection diagrams of both aged and unaged specimen are shown in the Fig. 4.

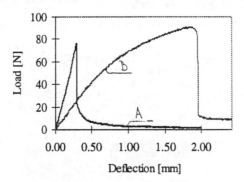

Figure 4 - Typical load-deflection diagrams of specimens subjected to A) accelerated ageing, B) normal ageing procedures .

The results of all three-point bending tests were statistically evaluated and are shown in Table 1.

	Max. load [N]	Deflection [mm]	Crack.load [N]	Deflection[mm]
Normal ageing	88.60 (5.35%)	1.91 (18.48%)	48.72 (13.74%)	0.57 (27.17%)
Accelerated ageing	69.70 (10.57%)	0.282 (8.88%)	N/A	N/A

Table 1 - Results of 3-point bending tests

The SEM photographs presented on Figs 5 and 6 show the difference observed in the failure modes of the specimens subjected to the accelerated ageing process.

Figure 5 - Typical failure mode of the bundle of fibres subjected to accelerated ageing process.

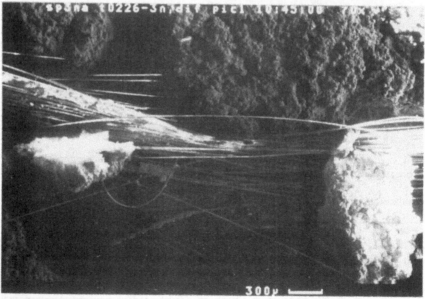

Figure 6 - Typical failure mode of the bundle of fibres subjected to normal ageing process.

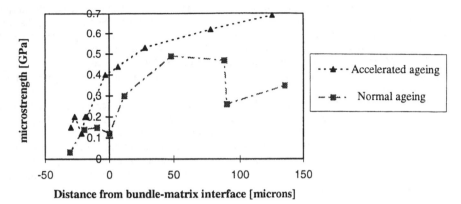

Distance from bundle-matrix interface [microns]

Figure 7 - Typical results of microstrength tests plotted against the distance from the bundle of fibres

DISCUSSION

SEM in situ bending test offers a viable alternative to previously used in-situ test methods. Since the specimens are subjected to bending and the load cell is connected directly to the data capture system it is possible to determine the load on the specimen at any moment during the testing. Another advantage of the test method presented lies in a much more precise determination of the point where a loading crack starts its propagation. The notches employed in the previous studies (Mindess, 1980; Bentur, 1984; Swift, 1987; Zhu, 1996) were all cut by a saw. The tip of such a notch had been usually at least 0.5mm wide. The tip of the V-shaped notch used in this project is only tens of microns wide. The start of the crack propagation is therefore identified much more precisely. Since the notches were produced after the ageing procedures had finished, no apparent shrinkage cracks were observed.

The major drawback of the current arrangement of the test method as presented is in the necessity of an adequate drying of the specimens before testing. Unfortunately, the authors did not have access to a "wet cell" SEM in which the testing and the observation of moist specimens has been proven to be practicable (Lovell, 1983).

The results of the microstrength tests provided additional interesting information about the material properties of the bundles of fibres itself and their interfacial transition zones. As it is shown in diagram on Fig. 7, the microstrength of ITZ decreases with decreasing distance between the centre of indentation and the outer interface of the bundle of fibres. However, the fact that the indents made were 15 microns deep should be taken into consideration for the results in interval from minus 20 to plus 20 microns. There the tip usually hits both the peripheral glass filaments from the interface of the bundle and ITZ.

CONCLUSIONS

At this stage, major outcomes of this investigation can be summarised as follows:

1. A new test method which provides opportunity for both qualitative and quantitative observations of microfracture mechanisms of composites has been developed.

2. Used in parallel with the microstrength and push-in tests, the SEM in-situ bending test provides a new tool for assessment of microfracture mechanisms of cement-based composites reinforced by fibres.

3. In comparison with previously used methods, the SEM in situ bending test provide the means for an examination of specimens of a more realistic size and reinforcement ratio. As the test simulates well the bending test, the specimens are subjected to a common, practical type of loading and there are no difficulties with mounting of specimens in the test stage.

4. The SEM in situ bending test offers new avenues for additional investigation e.g. measurements of crack mouth opening/displacement, observations of the crack path and its propagation etc.

5. The results obtained by means of the SEM in-situ three point bending test confirmed dissimilarities in performance of aged and unaged GRC. The failure modes of the bundles of fibres subjected to bending differed greatly according to the ageing procedure.

6. The microstrength test confirmed that there was a measurable increase in the microstrength value in the core of the bundle during the ageing process. This outcome offers an explanation for the change of failure mode of the bundle of fibres and hence for the deterioration of the bending performance of the aged material.

REFERENCES

Bartos, P. J. M. & Zhu, S. (1997). Assessment of interfacial microstructure and bond properties in aged GRC using a novel micro-indentation method. *Cement & Concrete Research,* **27** (1997) 1701-1712.

Bentur, A. & Diamond, S. (1984). Fracture of glass fiber reinforced cement. *Cement and Concrete Research,* 14 (1984) 31-42.

Duris, M. (1993). Micromechanics of fracture of inclined fibres in the cement-based composite. PhD thesis 252pp. University of Paisley, Paisley, Scotland, UK

Diamond, S. & Bentur, A. & Mindess, S. (1985). Cracking processes in steel fibre reinforced cement paste. *Cement & Concrete Research,* **15** (1985) 331-342

Lovell, J. & Diamond, S. & Mindess, S. (1983). Use of Robinson backscatter detector and "wet cell" for examination of wet cement paste and mortar specimens under load. *Cement & Concrete Research,* **13** (1983) 107-113

Marrs, D. L. & Bartos, P. J. M. (1996). Development and testing of self-compacting low strength slurries for SIFCON. In *"Production Methods and Workability of Concrete"* P.J.M. Bartos et. al Eds., E & FN Spon, Chapman &Hall, London, UK , (June 1996) 199-208

Mindess, S. & Diamond, S. (1980). A preliminary study of crack performance in mortar. *Cement and Concrete Research,* **10** (1980) 509-519

Swift, D. S. (1987). Bond in cement based composites reinforced with bundles of fibres. PhD thesis 293pp. Paisley College of Technology, Paisley, Scotland, UK

Tait, R. B. & Garrett, G. G. (1986). In situ double torsion fracture studies of cement mortar and cement paste inside a scanning electron microscope. *Cement & Concrete Research,* **16** (1996) 143-155

Zhu, W. (1995). Effect of ageing on durability and micro-fracture mechanisms of fibre reinforced cement composites. PhD thesis 273pp. University of Paisley, Paisley, Scotland, UK

PART SEVEN
ITZ STRUCTURE AND PROPERTIES IN SPECIAL CONCRETES

24 MICROHARDNESS AND MECHANICAL BEHAVIOR OF THE EXPANDED SHALE CONCRETE

S.-W. CHEN and U. SCHNEIDER
Institute of Buildings Materials, Building Physics and Fire Protection,
Vienna University of Science and Technology, Vienna, Austria

Abstract
The morphology of the interfacial zone between cement paste and expanded shale grains was investigated by means of scanning electron microscopy (SEM). The microhardness of the interfacial zone close to the dry or wet expanded shale aggregate surface was measured. Further compressive and flexural strength as well as elastic modulus of the concrete with dry or wet expanded shale were determined respectively.

The results show that the microhardness of the hardened cement paste about 20 μm near to the dry expanded shale grains is about two times that of 2000-4000 μm far from the grain surface. The contact area between the hardened cement paste and the wet expanded shale grains is weaker compared to the bulk of the hardened cement paste, similar to the situation of normal concrete. But the microhardness of the hardened cement paste increases with the distance from 60 μm up to 160 μm away from the interface. It was observed that the development of the compressive strength, flexural strength and elastic modulus of the concrete with wet expanded shale aggregate is faster than that with dry aggregate.
Keywords: Expanded shale; Interface; Lightweight aggregate; Lightweight aggregate concrete, Mechanical behavior of concrete, Microhardness, Transition zone.

The Interfacial Transition Zone in Cementitious Composites, edited by A. Katz, A. Bentur, M. Alexander and G. Arliguie. Published in 1998 by E & FN Spon, 11 New Fetter Lane, London EC4P 4EE, UK, ISBN: 0 419 24310 0

1 Introduction

It is known, cement concrete can be described as a multiphase composite in which different phases work together through the bonds of interfaces inbetween the phases. It can be assumed that during the whole life-time of concrete, from structural formation to structural damage and breakdown, inside the concrete there are permanent processes of interface-disappearance, interface-appearance and interface-transformation. In short, during the whole life-time of concrete the interfacial changes take place.

Concrete is simply described as a two-phase-composite construction material of coarse and fine aggregates in a matrix of cement paste. The interface zone between cement paste and aggregate in ordinary concrete is usually considered as a weak link in the concrete. The strength of concrete depends partly from the type of the cement-aggregate bond. Many studies concerning concrete microstructure, have concentrated on this aspect in the past [1][2][3][4].

In fact, the so-called interface of cement paste and aggregate in concrete is not a real geometrical plane but a transition layer (transition zone) with a specific microstructure and properties which differ not only from the aggregate but also from that of cement paste in bulk, and much of the paste in concrete or mortar is in this category [5][6]. According to *J. Grandet* and *J. P. Ollivier* [7] the $Ca(OH)_2$ crystals close to the interface indicates a strong preferred orientation, with the C-axis normal to the interface. The degree of orientation decreases with the distance and becomes zero at about 40 μm, i.e., the thicknesses of the transition layers between the cement paste and the limestone, quartz or polyethylene respectively, are all about 40 μm.

The study of the transition layers in lightweight concrete i.e. between cement paste and lightweight aggregate has not been documented very well to our knowledge. *J. K. Bhargava* has determined the interface of cement paste with expanded clay as well as pumice by means of radiography. The results indicated that the porosity of the hardened cement paste within the interfacial zone is about ±13.2% lower than that of the bulk cement paste. The thickness of the dense transition zone around expanded clay grains is inbetween 200-300 μm [8].

2 Microhardness of the transition zone

The Expanded Shale Grain (following for short ESG) roasted at high temperature has a rough surface and contains a lot of open pores and closed pore spaces. Optical microscopy and scanning electron microscopy were used to observe the interfacial layer between ESG and Hardened Cement Paste (following for short HCP).

The experimental results show that the cement paste was strongly adhered to the aggregate grains in the main cases under observation in spite of the presence of air bubbles and flaws. It was observed that the fresh cement paste (cement lime) was sucked into the open capillary pores of aggregate grains resulting in a significant strengthening of the mechanical bonds.

The microhardness of the hardened cement paste close to the expanded shale grains was measured by means of an automatic micro hardness tester which is a part of the polarization microscope "ORTHPLAN".

Dry expanded shale grains dried out at 105 °C and wet grains soaked in water for 24 hours were used as samples. The fresh cement paste with w/c=0.32 was cast into steel mould of 20x20x20 mm, one expanded shale grain was immersed into the fresh cement paste, and the mould was slightly vibrated by hand. Before the immersion the surfacial water of the wet aggregate was wiped by a wet cloth. After 24 hours the mould was removed and the cubic specimen was stored in a curing room at 20±3 °C and r.h.≥90%. The microhardness was measured at the ages of about 100 days.

For the microhardness test the orientation of the test plane was not considered due to above mentioned sample preparation. In order to get more data the measuring points were arranged as criss-cross. The microhardness of the lightweight aggregate were not determined due to its porosity and brittleness. The results of the mean microhardness value from over 200 test points of the 5 specimen is shown in Fig. 1. The variation of the microhardness values is up to 25%.

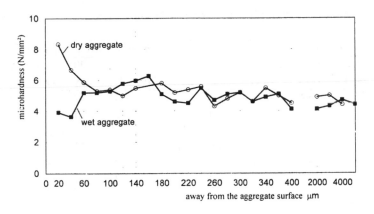

Fig. 1. The microhardness of the HCP around the ESG

The results indicate that the microhardness of the hardened cement paste about 20-40 μm to the dry grain surface is about two times higher than that of the 2000-4000 μm far from the surface. The average hardness value of the paste in the range up to 200 μm is 22.5% higher than that of the farther regions.

The contact area between the hardened cement paste and the saturated expanded shale grains is weaker compared to the neighbouring region, the microhardness of 40-60 μm from the interface is much lower than that of the bulk. It is of interest to note the microhardness of the HCP from 60 μm up to 160 μm away from the cement/aggregate interface. The mean value of the microhardness of the range of 60-160 μm is 1.9 times as high as that of the in bulk (2000-5000 μm).

3 Mechanical properties of the concrete

3.1 Compressive and flexural strength of the concrete

The dry expanded shale (moisture content about 3.6%) and wet one (soaked in water for 1 hour) were used for the preparation of the specimens. The mixtures with different net W/C-ratio were mixed in a laboratory concrete mixer, compacted on a vibrator table and cured at T=20±3°C and relative humidity r.h.≥.90%. The volume fraction of the expanded shale is 50% of the whole mixture. The total water comprises the net w/c with an addition of water due to the 1 hour water adsorption of the expanded shale for the mixture with the dry aggregate. The compressive strength of specimen of 100x100x100 mm and flexural strength of specimen of 65x65x240 mm were measured at different ages. The obtained results are listed in table 1 and table 2 (average value of three samples with a variation ≤ 15%).

Table 1. Compressive strength of the expanded shale concrete

Mixture proportions				Agg.	Compressive strength (N/mm²)					
W/C (Net)	Cement (kg)	Sand (kg)	E-shale (kg)		7 days	28 days	60 days	90 days	120 days	150 days
0.32	360	584	485	dry	22	34	37	37	---	36
				wet	24	29	34	33	---	38
0.38	360	584	508	dry	26	32	37	37	37	---
				wet	21	32	33	33	37	---
0.50	360	584	555	dry	21	24	21	29	31	---
				wet	17	27	28	27	34	---

Table 2. Flexural strength of the expanded shale concrete

Mixture proportions				Agg.	Flexural strength (N/mm²)				
W/C (Net)	Cement (kg)	Sand (kg)	E-shale (kg)		7 days	28 days	60 days	90 days	120 days
0.32	360	584	485	dry	2.4	3.0	3.0	2.9	2.8
				wet	2.1	2.9	2.7	2.7	3.1
0.38	360	584	508	dry	2.2	2.8	2.6	2.7	2.7
				wet	1.7	2.3	2.6	2.6	2.9
0.50	360	584	555	dry	1.8	2.4	2.7	2.6	2.2
				wet	1.6	2.1	2.6	2.6	2.1

The results indicate that the compressive strength and flexural strength of the concrete with the wet expanded shale were less than those of the concrete with the dried aggregates in the early age. But after hydration over three months the mechanical properties of the concrete with the wet expanded shale were nearly the same or more than those of the concrete with the dry aggregate.

For example, both the compressive strength and the flexural strength of the concrete with the wet expanded shale were about 10%-25% less than those of the dried one at an age of 7 days. The strength of the concrete with the wet aggregate increased significantly with time. At an age of 28 days the compressive strengths of the concrete (W/C=0.38 and 0.50) with the wet aggregate had nearly the same value than that of concrete with the dried aggregate. The compressive strength of the concrete (W/C=0.32) with the wet aggregate was about 14% less than that with the dried one at an age of 28 days, and more than that at an age 150 days. The mean flexural strength of the concrete with the wet aggregate was about 15% less than that with the dried one at an age of 7 days. And at an age of 90 days both types of concrete had nearly the same flexural strength.

3.2 Elastic modulus of the concrete

Under external load the elastic deformation of lightweight aggregate concrete depends not only on the short time deformability of the hardened cement paste and the aggregate but also on the volume fraction of the cement paste in the concrete. For constant volume fractions of the aggregate in the concrete the elastic deformability of the concrete depends only on the deformability of the hardened cement paste.

Table 3. Elastic modulus of the expanded shale concrete

Mixture proportions				Aggregate	Elastic modulus $(\times 10^3 \text{ N/mm}^2)$		
W/C (Net)	Cement (kg)	Sand (kg)	E-shale (kg)		7 days	28 days	90 days
0.32	360	584	485	dry	16.9	21.6	22.4
				wet	14.7	19.3	21.6
0.45	360	584	535	dry	13.9	17.7	19.3
				wet	12.0	16.9	19.6
0.50	360	584	555	dry	14.6	16.6	19.1
				wet	12.4	16.0	20.7

Table 3 shows the results of elastic modulus for the investigated expanded shale concretes with wet and with dried aggregates measured by the use of strain gauges. The preparation of specimens was the same as that of specimens for strength tests. The results show that after 7 days the growth rate of the elastic modulus of the concrete with the wet expanded shale was much higher than that of the dried one. The elastic modulus of the concrete with the wet aggregate at the age of 90 days increased by a factor of about 1.45 - 1.70, and that with the dried aggregate increased by a factor of about 1.30 - 1.40 in the whole range of the experiment. Therefore after about 90 days the elastic modulus of the concrete with the wet expanded shale overtook that with the dried aggregate.

4 Analysis and discussion

Expanded shale is an artificial building material which is roasted at high temperatures ranging from 1100-1250 °C. The active silica and active alumina on the surface of the expanded shale grains are able to combine chemically with the $Ca(OH)_2$ of the cement paste and form the contact points between aggregate and cement paste.

Characteristic for expanded shale grains is their rough surfaces and the open capillaries which not only enhance the mechanical bonds but also increase the actual contact area between the expanded shale grains and the cement paste, so that the physical adsorption and chemical bonding between the two phases are advanced.

The expanded shale grains are porous and suck parts of the mixing water from the fresh concrete mixture before the setting and hardening of the fresh concrete starts. This causes a reduction of the real W/C-ratio around the aggregate grains, i.e. its structure becomes closer and denser, in other words the expanded shale concrete has the effects of "self vacuumizing" and "self densifying". During the concrete hardening the sucked water within the aggregate grains is able to migrate back into the hardened cement paste which closes and densifies the grain surface due to temperature, humidity and capillary pressure differences between the aggregate grain and the hardened cement paste. That means, in the expanded shale concrete there are effects of "self moistening" and "self curing", so that the unhydrated cement grains in the hardened cement paste could further hydrate and the structure of the hardened cement paste could further be improved.

The effects that lightweight aggregate grains first suck mixing water from the concrete mixture , and later transfer it back into the cement paste are usually called the "micro-water pump-effect" of lightweight aggregate in the concrete. On the basis of our experimental results, we have found that this effect is a dominant reason for the development of a dense transition zone between the expanded shale and the cement paste. It is clear that for the dry expanded shale grain the "micro-water pump-effect" is powerful, and for the wet one the early effects of the "self vacuumizing" and the "self densifing" would be reduced but the later effects of the "self moistening" and the "self curing". This is the reason why the growth rate of the compressive strength, flexural strength and the E-modulus of the concrete with the wet expanded shale were much higher than those with the dried one, and why the microhardness of the hardened cement paste close to the wet expanded shale grain surface was comparatively strong.

The "micro-water pump-effect" of lightweight aggregate depends on the moisture and temperature of the aggregate, w/c-ratio of the concrete, curing conditions and the environment of the concrete in service. The sucked water within the aggregate grains can migrate back to the hardened cement paste, which depends simply on the differences of the temperature, humidity and compression between the aggregate grain and the hardened cement paste. The authors have found that the air locked in the aggregate pores by water had a dominant effect on the water migration.

5 Concluding remarks

In contrary to normal aggregate, lightweight aggregate (for example expanded shale or expanded clay) have a rough surface and porous structure which is beneficial to the bonding between cement paste and aggregate in the concrete. Especially the so-called "micro-water pump-effect" of lightweight aggregate grains in the concrete influences highly the strength of the bonds. This is beneficial for the interfacial zone and also for the wide-ranging surface bonds which are involved.

The contact zone between dried expanded shale grain and hardened cement paste is rather strong. The microhardness of the hardened cement paste about 20 μm near to the dried grain surface is by factor of 2 larger than that of the far distances of about 2000-4000 μm.

The contact zone of about 40-50 μm thickness between the wet expanded shale grain and the hardened cement paste is relatively weak. From 60 to 160 μm the microhardness of the hardened cement paste is much higher than that of the far distance from the cement/aggregate interface at an age of 100 days.

The growth of the compressive strength, flexural strength and the E-modulus of the expanded shale concrete with the wet aggregate are much higher than those with the dried aggregate, so that the mechanical properties of the wet aggregate concrete are the same or even higher than that of the dried one after curing for 3 months. This is due to the dense and hard bond between the cement paste and the aggregate, which develops around the dried expanded shale grains and also around the wet one during curing. In practice it is preferable that moistened lightweight aggregate grains are to be used for a concrete mixture, because this improves the workability of the mixture and also the microstructure of the transition zone of the aggregate/cement and therefore the mechanical properties of the entire concrete.

6 References

1. Diamond, S. (1986) in *8th ICCC*, vol.1, p.122.
2. Maso, J. C. (1980) in *7th ICCC*, vol.1, p.VII-1/3.
3. Massaza, F. and Costa, U. (1986) in *8th ICCC*, vol.1, p.158.
4. Struble, L., Skalny, J. and Mindess, S. (1980) in *Cem. Concr. Res.*10, 277.
5. Taylor, H. F. W. (1990) *Cement Chemistry.* ACADEMIC PRESS LIMITED, London.
6. Diamond, S., Mindess, S. and Lovell, J. (1982) in *Liaisons Pates de Ciment Materiaux Associes Proc.* RILEM Colloq, p. C42, Laboratoire de Genie Civil. Toulouse.
7. Grandet, J. and Ollivier, J. P. (1980) in *7th ICCC*, Vol. 3, pp.VII-63, VII-85.
8. Bhargava, Jitendra K. (1970) Radiographic Study of Paste Aggregate Interface in *Lightweight Concrete,* Royal Institute of Technology Stockholm, Sweden.
9. Farran, J. (1956) *Rev. Mater. Constr.* (490-491)155; (492)191.

10.   Farran, J., Javelas, R., Maso, J. and Perrin, B. (1972) *CR Acad. Sci.* Paris D275, 1467, Paris.

11.   Lea, F. M. (1970) *The Chemistry of Cement and Concrete (3rd Ed.),* Arnold, London.

12.   Hanson, J. A. (1968) American Practice in Proportioning Lightweight-Aggregate Concrete. in *1th Int. Cong. on Lightweight Concrete,* London.

13.   Struble, L. and Mindess, S. (1983) in *J. Cem. Comp. Lightwt Concr.* 5, 79.

14.   Monteiro, P. J. M., Maso, J. C. and Ollivier, J. P. (1985) in *Cem. Concr. Res.* 15, 953.

15.   Wu, Z.-W., Liu, B.-Y. and Xie, S.-S. (1982) in *Liaisons Pates de ciment Materiaux Associes* (Proc. RILEM Collog.), p.A28, Laboratoire de genie Civil, Toulouse.

16.   Glucklich, J. (1971) The Strength of Concrete as a Composite Material. in *1th Int. Conf. on Mechanical Behaviour of Materials,* vol. 4, pp.104-112.

17.   Harmathy, T. Z. (1967) *Moisture Sorption of Building Materials.* Ottawa.

18.   Wu, Z.-W. (1981) *The Theoretical Study on Cement and Concrete.* Technical University of Wuhan.

19.   Wu, Z.-W. (1980-1981) Materials Science of Concrete - the Composition and the Structure of Concrete. *Concrete and Structural Member,* Beijing.

20.   Igarashi, S., Bentur, A. and Mindess, S. (1996) Microhardness Testing of Cementitious Materials. *Advanced Cement Based Materials,* Vol. 4, Nr. 2, pp. 48-57,New York.

21.   Schneider, U. and Chen, S.-W. (1992) The Interface Zone around Expanded Shale Grain in Hardened Cement Paste.
       RILEM-Proceedings: *Interfaces in Cementitious Composites,* pp.149-156, E & FN SPON, London/New York.

25 MICROCEMENTS AS BUILDING BLOCKS IN THE MICROSTRUCTURE OF HIGH PERFORMANCE CONCRETES

D. ISRAEL
Dyckerhoff Zement GmbH, Wilhelm Dyckerhoff Institut, Wiesbaden, FRG
W. PERBIX
Dyckerhoff Baustoffsysteme GmbH, Wiesbaden, FRG

Abstract
Due to their special properties such as increased density and improved strength high performance concretes (HPC) are used in many different applications in civil engineering. Usually HPC is made of conventional cements modified with microfiller, for example silica fume. In the present paper, the use of microcements as a new way for modifying conventional cements for HPC is introduced. Microcements are special binders based on Portland cement or granulated blast furnace slag characterized by high fineness and limited maximum particle size compared to conventional cements. With the help of microstructural studies, it is shown how these microcements can be used as building blocks for the formation of a densified structure of the cementitious matrix and the interfacial transition zone (ITZ).

The suitability of composite binders made by mixing of conventional and properly chosen microcements to obtain HPC is shown with the example of a concrete under compressive load. The strengths of these concretes are increased up to 100% related to the reference.

Finally, further studies on the correlation of qualitative and quantitative properties of microcements with the microstructure of the resulting cement paste are discussed.
Keywords: High performance concrete, microcement, microstructure, model, particle size

1 Introduction

High performance concrete (HPC) can be characterized amongst other properties by its higher density, lower permeability, and higher strength, compared to ordinary concrete. These properties are the main reason for the increasing application of HPC in civil, building and hydraulic engineering. According to the kind of building these special

The Interfacial Transition Zone in Cementitious Composites, edited by A. Katz, A. Bentur, M. Alexander and G. Arliguie. Published in 1998 by E & FN Spon, 11 New Fetter Lane, London EC4P 4EE, UK, ISBN: 0 419 24310 0

properties are decisive features for the successful application. In Table 1, an overview of the properties of HPC related to their respective field of application is given [1].

Table 1. Favourable properties of HPC for different applications (leaf from [1])

	Strength	Durability	Freeze-thaw-resistance	Impermeability
Tower blocks	x			
Bridges	x	x	x	
Offshore buildings	x	x		
Tunnels		x		
Containers				x
Precast elements	x			x

It is known that the properties of concrete are not only determined by the qualitative and quantitative coordination of both the components, aggregate and cement paste, but crucially by the structure of the cement matrix and the interfacial transition zone (ITZ) between aggregate and cement paste. In standard concrete, this ITZ is the decisive weak point for premature failure of the concrete under mechanic and/or chemical load. Therefore, microcracks in overloaded standard concretes form mainly in the transition zone between cement and aggregate.   Consequently, an improvement of the microstructure of the concrete is particularly achieved by the reduction of the porosity of the hardened cement paste in the ITZ [2,3].

The quality of the hardened cement paste is basically determined by:

- the kind of cement
- the water/binder ratio
- the granulometry of the cement and
- the type of additives.

Usually, cements of normal composition and fineness are used as basic binders for the manufacture of HPC. In order to avoid macroporosity in the hardened cement, HPC is normally made with a water/binder ratio of less than 0.4. The optimization of the particle size distribution of the basic binder is achieved by the addition of additives. In the past, mainly microsilica and special fly ashes have been used as suitable microfillers for the densification of the microstructure of concretes.

In the present paper ways to modify the granulometry of the basic binder by the use of specifically selected microcements are shown. With the specific coordination of the particle size of the conventional cement and the microcement, which possibly is combined with silica fume, it is aimed at a high packing density of the cement particles. Additionally, the control of the hydration process becomes possible due to the coordination of the chemical composition of both binders.

2 Specification of the microcements

Microcements are materials of high fineness, based on Portland cement clinker and/or special granulated blast furnace slag. In order to guarantee regular properties of these binders, their components are preferentially produced by separate milling of the mineral raw materials, the separation of their finest fractions and their putting together according to the formulation of the respective microcement. As cement-like raw materials are used the chemical composition of the microcements is comparable to conventional cements [4].

The most important characteristic of microcements distinguishing them from conventional cements is the high fineness and simultaneous limitation of their largest grain size. A comparison of particles of a very fine microcement and an ordinary CEM I 52.5 R is given in Figure 1. The maximum particle size (d_{95}) of the shown cement is approximately 30 μm, whereas the d_{95} of the shown microcement is approximately 6 μm. For an estimation of the ratio between the maximum particle sizes of a conventional cement and microcements a factor of 5 can be used.

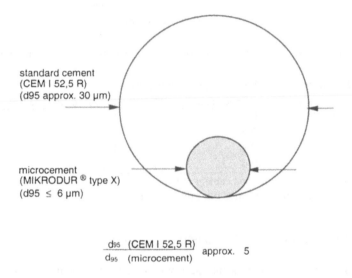

$$\frac{d_{95} \ (CEM\ I\ 52{,}5\ R)}{d_{95} \ (microcement)} \quad approx. \ 5$$

Figure 1. Comparison of maximum particle size of conventional cement and microcement

The limitation of the particle size in the particle size distribution of the binders is causing a clear shift of the grading curves of the microcements with regard to the conventional cements to smaller particle diameters. In order to distinguish microcements from conventional cements the grading curves of the currently available microcements (maximum particle size from 6 and 24 μm, resp.) are compared in figure 2 to representative grading curves of two conventional cements (CEM I 32.5 R and CEM I 52.5 R, resp.).

It can be seen from these graphs (Figures 1 and 2), that by proper choice of the particle sizes of basic binder and microcement, the granulometry of the composite binder can be optimized in a way that the microcement particles serve as building blocks for filling the hollow spaces between the coarser particles of the conventional cement. This is achieved by the provision of microcements with defined grain size distribution and fixed maximum particle size (Table 2).

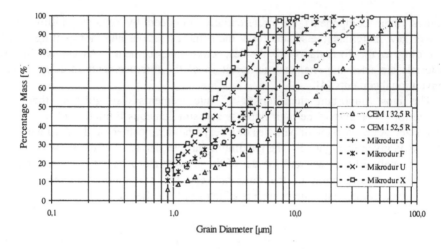

Figure 2. Particle size distribution of conventional cements and microcements

Table 2. Specification of MIKRODUR® microcements

max. particle size (d₉₅)	microcement type P*	microcement type R**
≤ 24 µm	P-S	R-S
≤ 16 µm	P-F	R-F
≤ 9.5 µm	P-U	R-U
≤ 6 µm	P-X	R-X

* microcement based on Portland cement clinker
** microcement based on granulated blast furnace slag

3 Microstructural model of HPC modified by microcements

A major parameter determining the quality of hardened cement paste is its porosity. This porosity is caused by the hollow spaces between the cement particles, which are existent even with optimal packing, as well as by the pores due to the excess water in the cement paste. Especially the latter effect is determining the microstructure of the transition zone between cement matrix and aggregate. This is caused by the fact that in the fresh concrete the aggregate grains are covered by a water film resulting in the local increase of the water/cement ratio. Thus, favourable conditions for the preferential formation of plate-like calcium hydroxide crystallites are created, which impair the quality of the bonding system aggregate/cement (Figure 3).

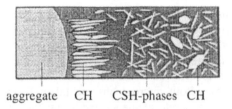

aggregate   CH   CSH-phases   CH

Figure 3. Illustration of the interfacial transition zone
between aggregate and cement matrix [5]

By modifying conventional cements with microcements, the cement matrix is densified
and the bond between aggregate and cement is improved. This happens mainly
because of the homogeneity of the microstructure due to the reduction of pore volume
by the filling of the hollow spaces with finest particles (filling effect, Figure 4). The
denser physical packing of particles results in more reactive grains per volume than
what could be achieved without reactive microcements. The wetting of the surfaces of
all solids is supported by the addition of plasticizers.

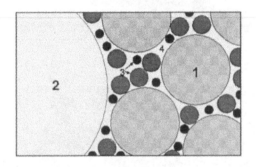

1 conventional cement   2 aggregate   3 microcement   4 water
Figure 4. Illustration of the ITZ of a composite binder

The enlargement of the specific surface causes a higher chemical reactivity of the
composite binder, which togehter with the introduction of a multitude of nuclei results
in an accelerated cement hydration. Additionally, the similar composition of
microcements and conventional cements supports this acceleration. Caused by the
limited space the hydration products are hindered in their growth, and thus small, well
linked calcium silicate hydrate phases are forming [5]. These effects result in a
decreased porosity in the bulk of the cement matrix as well as in the ITZ between
matrix and aggregate. If microcements with latent hydraulic properties on the base of
granulated blast furnace slag are used the characteristic consumption of calcium
hydroxide occurs during setting and the amount of the unfavourable calcium hydroxide
is reduced. Furthermore, the positive impact of finest additives on the dimension of
the ITZ is known [6,7,8].

Figure 5 verifies the changes of the microstructure, which are attributed to the use of microcements in the manufacture of HPC and are described with the help of the model. The porous microstructure consisting of calcium silicate hydrate phases and calcium hydroxide crystalls in the area of the ITZ of a conventional cement can be seen in Figure 5a, whereas this area is much denser in the case of a HPC modified with microcement (Figure 5b). Both micrographs are taken from fractured surfaces of specimens prepared from concrete elements and show the vicinity of an aggregate grain.

Figure 5. SEM micrographs of the ITZ in a conventional cement (top)
and a composite binder (bottom); scale bar 5 μm

4 Mechanical studies

The possibility of manufacturing HPC by using microcements is described in the following with the example of a concrete for the production of precast elements. The aim of the studies reported here was the improvement of compressive strength of a conventional concrete by the modification of the conventional cement (CEM I 42.5 R) with granulated blast furnace slag containing microcement and possibly silica fume.

Information on the formulation of the conventional concrete (reference) and the concretes prepared with modified formulations (HPC) is given in table 3. It was found that for the modification of the binder a combination of 10w% granulated blast furnace slag containing microcement and of 5w% silica fume was most suitable.

Table 3. Formulations of the tested concretes

Formulation	conventional concrete	modified concretes	
		(1)	(2)
Basic binder	CEM I 42.5 R	CEM I 42.5 R	CEM I 42.5 R
Composite binder	100 % CEM I 42.5 R (Reference)	85 w% CEM I 42.5 R 10 w% MIKRODUR® 5 w% silica fume	85 w% CEM I 42.5 R 10 w% MIKRODUR® 5 w% silica fume
W/C-ratio	0.5	0.5	0.36
Plasticizer	--	--	1.2 w%

In figure 6 the compressive strengths of the concretes with composite binders related to the reference concrete are plotted.

The effect of binder modification on the level of strength is clearly visible. Principally, the strength of the HPC is increased by up to 30% compared to the conventional concrete, no matter what age the concretes have. If simultaneously the water/binder ratio is reduced from 0.5 to < 0.4, which is typical value for HPC, the one-day compressive strenth is increased by a factor of 2.1. It is striking that the acceleration of the strength development is especially effective, if an additional decrease of the pore volume is achieved by reduction of the water/cement ratio. At later ages (2 and 28 days) the differences in strength are not as high, since the impact on reactivity at these ages is obviously of secondary significance. Nevertheless, an increase in strength of 60% and 50%, respectively, can be observed here.

The significantly higher early strength may give first hints on opportunites for the optimization of process technology of precast concrete elements, whereas the mentioned increase in strength of 50% is essential for the application of these HPC elements.

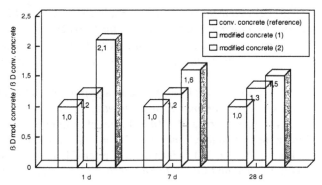

Figure 6. Ratio of compressive strengths of concretes made with conventional cement and composite binders

5 Further optimization of HPC modified with microcements

It is shown in the studies presented here, with the example of concretes under compressive load, that it is possible to make HPC with microcements. It can be assumed that the changes in the microstructure achieved by the application of microcements are reflected in important properties of HPC, such as improved freeze-thaw resistance, increased durability, and decreased permeability.

In current studies, existing computer models are adapted for the optimization of the packing density of composite binders used for the manufacture of HPC. These investigations are aimed at coordinating the particle size distributions of the microcements with those of the basic binders with the help of modelled microstructures. For that purpose the currently available microcements characterized by different granulometries are used like building blocks. A further possibility for optimizing the quality of HPC is to use information from the modelling of the hydration process on which microcement is used best for a special application.

From the presented results and the described models, it seems to be possible to create HPC on the base of conventional cements and properly chosen microcements according to specific demands. These composite binders can be designed by coordination of the characteristic composition and characteristic granulometry of both binders.

6 References

1. König, G., Grimm, R. (1996) Hochleistungsbeton, in *Betonkalender Teil II*, (ed. J. Eibel), Ernst & Sohn, Berlin, pp. 441-546.
2. Lange, F., Mörtel, H., Rudert, V., Sturm, I. (1997) Influence of microstructure on the durability of cement mortars. *Concrete Precasting Plant and Technology*, Issue 8, pp. 108-114.
3. Rudert, V., Strunge, J., Wihler, H.-D. (1994) Concrete from a different point of view - Filigree microstructure. *Concrete Precast Plant and Technology,* Issue 9, pp. 86-93.
4. Perbix, W. (1994) Feinstzemente für Injektionen, in *Proceedings, 12. Internationale Baustofftagung*, Weimar, Vol. 2, pp.119-128.
5. Lange, F. (1996) Gefügeuntersuchungen und Eigenschaften von Hüttensand enthaltenden Zementen, *Dissertation, Technische Fakultät der Universität Erlangen-Nürnberg.*
6. Lange, F., Mörtel, H., Rudert V. (1996) Properties of mortars containing fine-ground cements. *American Ceramic Society Bulletin,* pp. 107-09.
7. Detwiler, R., Krishnan, K., Metha, P. (1987) Effect of granulated blast furnace slag on the transition zone in concrete. *American Concrete Institute*, ACT-SP 100.
8. Goldman, A., Bentur, A. (1993) The influence of microfillers on enhancement of concrete strength. *Cement and Concrete Research*, Vol. 23, pp. 962-972.

26 BOND PROPERTY STUDIES ON CERAMIC TILE FINISHES

Z.J. LI, W. YAO, M. QI, S. LEE, X.S. LI, and C.H. LEE
The Hong Kong University of Science and Technology, Clear Water Bay,
Kowloon, Hong Kong, China

Abstract

To deal with the tile debonding problems in ceramic tile finishes, it is necessary to study the bond properties of ceramic tile systems. For this purpose, bond tests have been conducted using a newly developed push-off method in this study. The specimens were made by affixing a ceramic tile with the dimension of 150 X 150 mm to a surface of concrete cubes. Various parameters such as workmanship, temperature, and bonding materials have been investigated on their influences to ceramic bond properties. The observations from these preliminary test results are highlighted below: First, the pressure is very important to the development of bond between tile and render. Tiles fixed by trowel knocking (higher pressure) have a much stronger bond (about ten times) than tiles affixed by hand pressure (lower pressure). Second, pre-filling of the zigzag area on the back of a tile before attaching it on render has a significant influence on bond development. A forty percent increase in bond strength is observed for this kind of specimen as compared to specimens without pre-filling. Third, the cycling temperature has a smaller influence on bond strength of normal mortar and has no influence on the silica fume modified mortar. Fourth, incorporation of silica fume into mortar does increase the bond strength significantly, about eighty percent increase in bond strength can be achieved. Lastly, it seems that weak bond strength tends to cause a failure at tile-render interface (specimens with normal mortar) and strong bond tends to cause a failure at tender-concrete interface (specimens with silica fume modified mortar).
Keywords: Ceramic tile, bond properties, push off test, pressure, silica fume.

The Interfacial Transition Zone in Cementitious Composites, edited by A. Katz, A. Bentur, M. Alexander and G. Arliguie. Published in 1998 by E & FN Spon, 11 New Fetter Lane, London EC4P 4EE, UK, ISBN: 0 419 24310 0

1. Introduction

The decoration of high-rise buildings using ceramic tile finishes contribute not only to improve the stark appearance, but also offer some degree of protection from carbonation attack to the concrete surface beneath. Unfortunately, almost every building would experience some degree of failures on its ceramic tile finish [1]. This directly influence the effect of protection on the building. Most of these failures manifest themselves as hollowing, cracks, discoloration or disintegration of the finishing material, or by the finishing material separating from its support system. However, according to the statistics, it is concluded from these failure patterns that the hollow area of tile is the main cause of the tiling system defects, resulting in debonding and, in some instances, detachment and collapse.

Debonding is a common and an inevitable occurrence of tile finishes. Repairs for the debonded tiles are difficult and often require extensive unsightly scaffolding. In addition, noisy hammering and chipping to remove the offending material and prepare the substrate is accompanied by extensive dust and debris, which pollutes the environment and causes inconvenience for the residents.

To overcome these problems, it is necessary to improve the performance of ceramic tiling systems. Effort must be made to minimize or eliminate the hollow areas, the most important points are to improve the interface adhesive strength and modify finishing methods.

In order to optimize the bonding properties, experimental studies were carried out with various kinds of adhesive materials and workmanship. The purpose of this study is to provide some general guideline for the construction of a tile finish.

2. Experiments

2.1 Preparation of specimens

To prepare the push-off specimens, concrete cubes of 150 x 150 x 150 mm were cast first. The specimens for push off test were made by affixing a ceramic tile with the dimension of 150 X 150 x 8 mm to one surface of the cubes with different substrate materials. The mortar matrix for control specimen group was prepared by using Type I ordinary Portland cement and river sand. The mix ratio for the mortar is 1:0.5:3 (Cement: Water : Sand) by weight. The affixing method for control specimen is hand pressure with twist which can be frequently observed on a construction site. Besides of the control group, six groups of specimens using different matrix or affixing methods were prepared for the research. The details of the specimens used are listed in the Table I. For silica fume modified matrix (group F and G), 10% (by weight) of cement was replaced by silica fume. The specimens cured at a normal temperature were finished in a water tank under room temperature for 26 days after demolding, while specimens prepared at the cycling temperature, were cured in a water tank at 20°C for 18 days after demolding and then were put into an environmental chamber to undergo an accelerated curing with 50 fixed temperature cycles (from 0°C to 50° C) for 7 days. One day before testing, specimens were taken out of either water tank or curing chamber. A L-shaped steel plate was attached to the tile using epoxy resin. One day curing was required for these specimens before test.

Table 1. Type of specimens

Type	Adhesive materials	Pressure	Finish method	Curing condition
A	normal	hand	standard	constant temperature
B	normal	trowel knock	standard	constant temperature
C	normal	trowel knock	standard	cycling temperature
D	normal	trowel knock	modified	constant temperature
E	normal	trowel knock	modified	cycling temperature
F	modified	trowel knock	standard	constant temperature
G	modified	trowel knock	standard	cycling temperature

Note:
Normal materials   =cement and sand(1:3)
Modified materials = some part(10%) of the cement is replaced with silica fume.
Hand pressure   = hand pressure with twist is applied.
Standard method   = tile is fixed according to Clause 18.107 of H.K.ASD specifications [2].
Modified method   = the back of tile is coated with cement slurry prior to fixing tile on mortar.
Constant temperature = curing temperature is constant.
Cycling temperature = cycling 50 times between the lowest T(0∘C)and the highest T(50∘C).

2.2 Push-off test

The bond properties of interfaces between ceramic tile and substrates were measured using a push-off test conducted with a 25 metric ton MTS machine. The test setup is shown in Fig.1. As can be seen from the figure, the specimen was tied on a flat rectangular plate that was connected to a servo hydraulic actuator of a testing machine through two C-clamps. The entire specimen/fixture could move up with the actuator. A steel rod, contacted to the load cell, made contact with the top surface of the L-shaped steel plate with a help of a sphere joint. Restraining force provided by resisting upward movement of the loading fixture pushed the L-shaped steel plate, which in turn pushed the ceramic tile, downwards. Two LVDTs, which were fixed between the top surface of the specimen and the rigid wing of the steel rod, were used to measure the displacement at the top of the L-shaped steel plate relative to the surface of the concrete cube. The average output of the LVDTs was also used as a feed back signal to control the movement of servo hydraulic system. The rate of

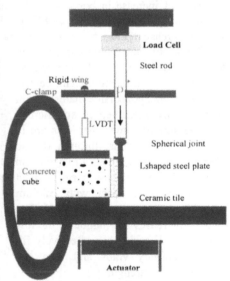

Figure 1 Push-out set-up

displacement of the push-out test was set at the range of 0.006-0.06 mm per minute. Push-off load, displacements of the tile and stroke of the actuator were recorded using a data acquisition computer. For each group of experiment, three specimens were tested. The bond properties for interface were obtained by averaging the three specimen results.

2.3 Test results

For a typical push off test, three measurements of importance are time, push off load, average displacement at the top of the specimen. For a simple analysis of the results, a graph of push off load vs. top displacement is produced as shown in Fig. 2. Taken from this graph is the peak load. By dividing the peak load by the original bonding area of a ceramic tile, the average bond strength can be obtained. Although this is a quite simple interpretation for the test results, it can be utilized to compare the influences of different parameters. The calculated results is shown in Fig. 3. From this figure, the following phenomena can be observed.

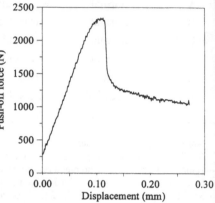

Figure 2 Push-off vs. displacement

(a) Influence of affixing pressure: As mentioned earlier, specimens in group A were made by hand pressure only. Obviously, the average bond strength of this group is much lower than that of all other groups. It can be found from the figure that the bond strength of Group B is about 10 times of that group A although the only difference between these two groups is the pressure for mounting tiles.

(b) Influence of back coating: For Group D, pre-filling of the zigzag area on the back of a tile before attaching it on render has a significant influence on bond development. A forty percent increase in bond strength was observed for this kind of specimen as compared to specimens without pre-filling. This was due to the reduction of the hollow area and increase of the actual adhesive area.

(c) Influence of silica fume: For Group F, adding silica fume (SF) to the mortar increased the bond strength significantly, SF is a very fine pozzolanic material with a particle size measured in submicrons while the size of cement particles are in tens of microns. Thus, through the so called packing effect, a denser material can be made. Also, as we know, during cement hydration. SF can react with calcium hydroxide and high calcium C-S-H. The reaction increases the amount of low calcium C-S-H which can effectively fill out the capillary voids and thus can improve mortar strength.

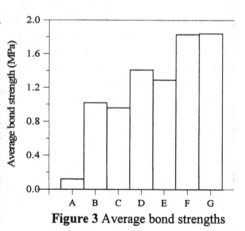

Figure 3 Average bond strengths

(d) Influence of temperature cycling: For Group C, E, G , it is found that the cycling temperature has a smaller influence on bond strength of normal material, and has no influence on the silica fume modified mortar because of the improvement in its internal microstructure.

2.4 Failure modes

There are two main failure modes during tests according to the adhesive strength. One is the debonding of the interface between the tile and the adhesive mortar (noted as 1/2) and another is the debonding at the interface between the adhesive mortar and the concrete (noted as 2/3). It seems that a weak bond strength tends to cause a failure at the tile-render interface(specimens with normal mortar) and strong bond tends to cause a failure at render-concrete interface(specimens with silica fume modified mortar). The statistical results of specimen failure are present in Table 2.

Table. 2   Descriptions of Specimen Failure Modes

Specimen remarks	Average failure force (N)	mode (Type: Est.%) Failure
A	2744	1/2:100%
B	23014	1/2:100%
C	21572	1/2:95%, 2:5%
D	31683	1/2:100%
E	28934	1/2:60%, 2:5%, 2/3:35%
F	41249	1/2:60%, 2:5%, 2/3:30%, 3:5%
G	41487	1/2:40%, 2/3:60%

3. Conclusions

a. The debonding of tiles is caused by two main factors: a bigger hollow area and lower adhesive strength. To avoid debonding, the adhesive materials and workmanship have to be improved.

b. Tiles fixed by trowel knocking have a higher bond strength than tiles affixed by using hand pressure.

c. Pre-filling of the zigzag area on the back of a tile before mounting it on render is good for bond development.

d. The addition of silica fume to the mortar increased the bond strength significantly. Eighty percent increase in bond strength can be achieved.

e. The cycling temperature has a smaller influence on the bonding strength of normal mortar and has no influence on the silica fume modified mortar.

4. Acknowledgement

The financial support from Architectural Service Department of Hong Kong Government is greatly acknowledged. Mr. A. R. Wilson and Mr. W. M. Tang are acknowledged for their assistance and advises to this project.

5. References

1.   H.Leslie Simmons.(1990). Repairing and Extending Finishes.   Van Nostrand Reinhold, New York.
2.   ArchDS Specification. HongKong. 1990.

PART EIGHT
EFFECT OF CHEMICAL AND PHYSICAL PROPERTIES OF AGGREGATES ON ITZ

27 INTERACTION BETWEEN FLY ASH AGGREGATE WITH CEMENT PASTE MATRIX

R. WASSERMAN and A. BENTUR
National Building Research Institute, Faculty of Civil Engineering,
Technion, Israel Institute of Technology, Israel

Abstracts
The interactions between fly ash aggregates and portland cement matrix was studied to determine factors other then aggregate strength which influence the concrete strength. The structure and properties of sintered fly ash lightweight aggregate was modified by heat and polymer treatments to obtain aggregates of variable properties. Concretes of equal effective water/cement ratio prepared from these aggregates were tested for strength and microstructure. It was found that the differences in the aggregate strength could not always be accounted for by the differences in the aggregate strength. The physical and chemical interfacial processes affect the overall strength beyond that of the aggregate strength. The physical process identified was densification of the interfacial transition zone at early age due to absorption of the aggregates. The chemical processes were associated with pozzolanic activity of the aggregate and deposition of CH in the shell of the aggregate. Such effects should be taken into account when predicting the concrete strength or in the design of lightweight aggregate of optimal properties.
Keywords: Aggregates, cement paste, concrete strength, fly ash, lightweight.

1 Introduction

In recent years there has been increased the application of lightweight aggregate for production of high strength concretes. Therefore there is renewed interest in the role of aggregates in such concretes, particularly since they are the weak link in these systems[1-8]. The mix design concepts are usually based on the production of high strength matrix of low water/cement ratio to compensate for the aggregate weakness. Recent studies into the microstructure of lightweight aggregate concretes suggest that the interaction of the paste matrix and the aggregate can be quite different from that of normal aggregate concretes. The interfacial microstructure can be quite similar to that of normal concrete if the aggregates are wetted. But it can be considerably denser if the aggregate is used dry. The influence of the chemical activity of the aggregate on the

The Interfacial Transition Zone in Cementitious Composites, edited by A. Katz, A. Bentur,
M. Alexander and G. Arliguie. Published in 1998 by E & FN Spon, 11 New Fetter Lane,
London EC4P 4EE, UK, ISBN: 0 419 24310 0

structure of the interfacial zone is not resolved as in some studies it has been suggested to have some influences[4,5], while in others no such interaction was observed[3,7].

In view of this significance of the aggregate matrix interactions in controlling the strength of high strength lightweight concretes, a systematic studies was undertaken with sintered fly ash lightweight aggregates to resolve some of the mechanisms by which they interact with the matrix, and assess their potential influence on the strength of the concrete.

2 Experimental

Commercial Lytag (UK) aggregate with uniform size grading (maximum size of 6 mm and fineness modulus of 4.1) was modified by 4 types of heat treatment (heating 1200, 1250 and 1300°C and rapid and slow cooling afterwards) and one type of polymer treatment (partial impregnation and hydrophobization) as outlined in Table 1, to obtain aggregates with controlled differences in absorption and pozzolanic activity .

Table 1. Treatments and Notation of Aggregates

Lytag	Untreated original aggregate
L1200	Heated to 1200°C, cooled rapidly in air
L1250	Heated to 1250°C, cooled rapidly in air
L1300SC	Heated to 1300°C, cooled slowly in oven
polymer	treated with polymer

Details of estimating for the mechanical quality of the aggregates as well as of concretes' preparation and curing are given in Ref. 9.

The concretes were tested periodically for compressive strength and their internal structure was studied by SEM observations at one day and at a later age of 90 days and micro-chemical analysis of polished surfaces. Energy dispersive X-ray analysis was used to determine the Ca and Si contents, and profiles of C/S ratio across the aggregate-paste interface were drawn.

3 Results and discussion

Table 2 summarizes main characteristics of the initial and treated aggregates. The changes in the properties of the aggregates due to the heat and polymer treatment are discussed in Ref. 9.

3.1 SEM observation
SEM observation at high magnification (Fig. 1) shows that at one day there seemed to be intimate contact between the aggregate and the matrix only in the concrete from the original Lytag aggregate.

Table 2. Properties of lightweight aggregates

Name	Bulk density (kg/m^3)	Density of solid (kg/m^3)	Mercury measured porosity (%Vol.)	Total water absorption (%Vol.)	Water absorption (%Vol.)	90 Days pozzolanic activity of the outer shell	Crushing Strength (MPa)
Lytag	1560	2470	37.32	19.14	14.50	0.15	14.9
L1200	1550	2400	36.20	14.11	10.60	0.28	19.3
L1250	1680	2340	27.91	12.40	9.70	0.40	14.7
L1300SC	1600	2340	32.57	10.62	9.20	0.52	15.6
polymer	1610	2350	30.55	8.42	3.80	-	19.9

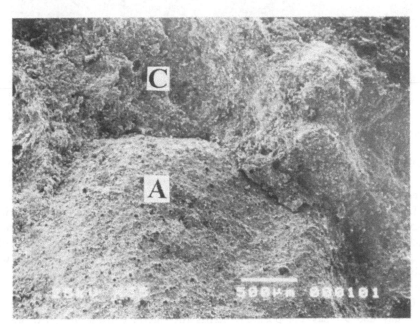

Fig. 1. The aggregate-matrix interface at one day: A - untreated Lytag; C - cement matrix

In all the treated aggregates separation was observed. This separation could reflect the presence of a very porous material which shrinks considerably during the drying in the preparation of the specimens for the SEM observation, with eventual separation. SEM observations of the treated aggregates, as well as the observations at the paste surface exposed after the aggregate removal are shown in Ref. 11.

The concrete with the initial Lytag aggregate is characterized by a denser microstructure. The paste with the treated aggregates are much more porous. In Fig. 2 it could be detected more needle-like material, evidently ettringite. At 90 days all the concretes with the heat treated aggregates seem to have a closed contact at the

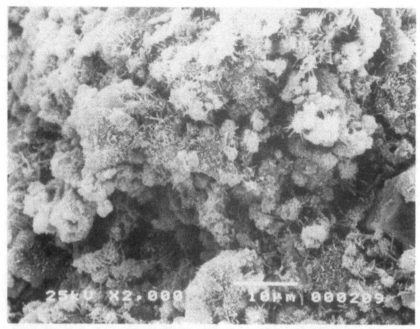

Fig. 2. View of the matrix side of the aggregate-matrix interface in the concrete with the heat treated aggregate (1300°C)

interface. The polymer treated aggregate concrete showed separation even at 90 days (Fig. 3). The surface of the original Lytag and heat treated aggregates is covered with cement paste matrix without being able to detect the porous nature of aggregates (Fig. 4). In comparison to these aggregates, kept the polymer treated aggregate its original obviously porous view. In contrast to polymer treated aggregate, in the case of the original Lytag aggregate there is evidence for presence of large crystals in the pores (Fig. 5) whereas in the heat treated aggregates no clear crystalline morphology could be observed.

3.5 Composition of the interfacial zone
The 1 day and 90 day compositions across the interface from the paste into the aggregates was determined in terms of C/S ratio. The C/S ratio of the aggregate shell itself is about 0.15 and that of the bulk paste is about 3. It was found that in the vicinity of the actual interface there are deviations from these bulk values which extend into the aggregate (Figs 6 and 7). The width of the interfacial transition zone in front of the aggregate can be estimated as the zone in which the C/S ratio is high, and it about 4 and 40 μm for the original Lytag aggregate and the 1300°C treated ones, respectively. The lower value for the untreated aggregate may be accounted for by the higher absorption which prevents accumulation of water at the interface: accumulation of such water in the fresh concrete is believed to be the cause for the formation of the porous zone in the vicinity of the aggregate surface, resulting in a more porous microstructure which is rich in CH. This trend changes at 90 days, where a high C/S ratio curve seems to be "penetrating" deep into the aggregate in the untreated Lytag aggregate (Fig. 6),

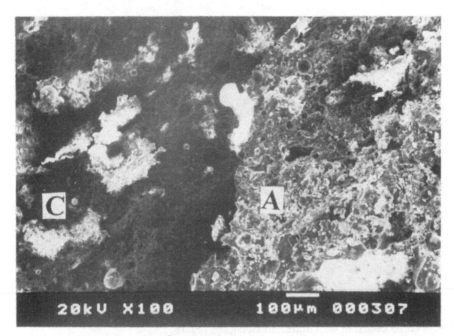

Fig. 3. The aggregate-matrix interface at 90 days: polymer treated aggregate

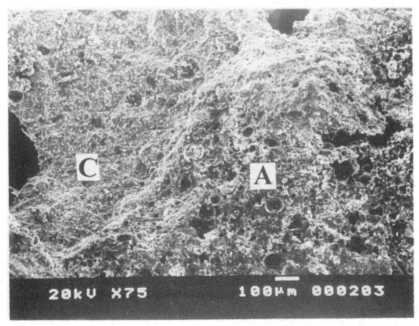

Fig. 4. View of the aggregate and of the aggregate-matrix interface at 90 days: A - untreated Lytag; C - cement matrix

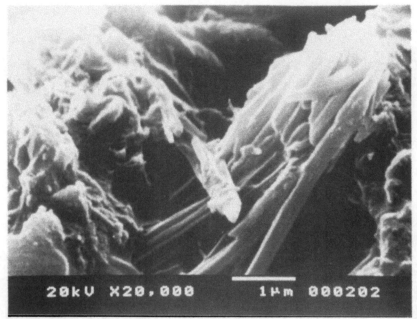

Fig. 5. High magnification of the aggregate surface at the aggregate-matrix interface at 90 days cured concrete with untreated Lytag aggregate.

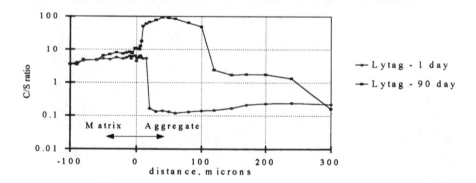

Fig. 6. C/S ratio profiles across the aggregate-matrix interface in 1 and 90-day-old concretes with untreated Lytag aggregate

whereas in the heat treated aggregates, at 1300°C, the C/S ratio curve is declining gradually within the aggregate boundaries, from values of about 3 at the interface to the characteristic aggregate C/S ratio of about 0.15. This occurs at a distance of about several tens of microns into the aggregate (Fig. 7). These results suggest that in the original Lytag aggregate the pore solution can penetrate effectively into the pores of the capacity, and deposits of CH are formed in the aggregate pores, as evidenced by the high C/S ratio and the SEM observations. In the higher temperature treated aggregate there is probably absorption of this kind, although to a smaller extent, and since the aggregates are more pozzolanic. CSH rather than CH is formed preferentially.

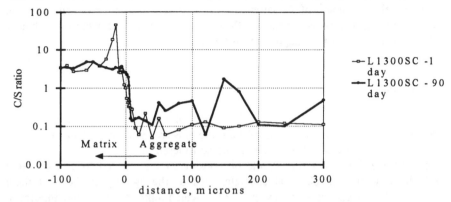

Fig. 7. C/S ratio profiles across the aggregate-matrix interface in 1 and 90-day-old concretes with L1300SC aggregate

3.6 Strength

The values of compressive strength of concretes prepared with the original and treated aggregates are shown in Table 3. The data is grouped into two classes of concretes each made of aggregates of similar crushing strength. It can be seen that comparison within each class shows that the strength values are not necessarily the same even though the aggregates and the matrix are of the same quality.

Table 3. Compressive strength of lightweight aggregate concrete

Group	Aggregate type	Aggregate Crushing Strength (MPa)	Concrete compressive strength (MPa) at:				
			1 day	7 days	28 days	56 days	90 days
	Lytag	14.9	10.5	38.8	45.4	53.6	57.7
I	L1250	14.7	9.3	34.6	41.0	52.8	57.5
	L1300SC	15.6	8.2	34.5	40.5	54.5	58.5
II	L1200	19.3	12.1	46.9	54.8	57.8	60.9
	polymer	19.9	9.8	35.7	43.6	45.0	45.3

For example, concretes with L1200 and polymer treated aggregates are quite different in their strength at all ages in spite of the fact that the two aggregates are of similar strength.

In group I the trends change with time: at early age the strength of the concrete is smaller for the higher temperature treated aggregate, but with time the difference diminish, and by 90 days they are practically equal. When comparing between the groups it can be seen that the higher strength heat treated aggregate in group II provides a concrete which at early age is stronger than all concretes in group I which is in agreement with the higher strength of the L1200 aggregate in group II.   However, this difference diminishes with time and by 90 days the concretes of group I are practically of the same strength as the L1200 aggregate concrete in group II, in spite of the higher crushing strength of the latter aggregate. The polymer treated aggregate in group II always gave a weaker concrete than those in group I, in spite of its higher strength compared to all of the aggregates in group I.

4 Conclusions

(1) Lightweight aggregates of similar strength do not necessarily yield concretes of equal strength even if the matrix is of the same effective w/c ratio. Thus, factors in addition to the aggregate strength should be considered.
(2) The additional influences that should be considered were found to be associated with the physical and chemical characteristics of the aggregates, both of them affecting the overall strength by processes which take place at the interfacial transition zone. In the aggregates this zone extended also into the aggregate itself.
(3) The physical process occurs at early age and is governed by the absorption of water into the aggregate. Higher absorption eliminates accumulation of water in the fresh matrix in the vicinity of the aggregate. As a result the interfacial transition zone in lightweight aggregates of higher absorption is denser. Thus, for lightweight aggregate aggregates of equal strength, the aggregate of higher absorption will provide higher strength concrete due to its denser interfacial zone.
(4) The chemical process occurs at later age. Two types of processes were resolved here: pozzolanic reaction between the aggregate and the alkaline pore solution which penetrates into it, and an "impregnation" process in which CH deposits in the pores of the aggregates. The latter is more likely to occur in aggregate having higher absorption and bigger pores.
(5) This is a basis for considering these mechanisms for controlled production of lightweight aggregates intended for use in high strength lightweight concrete.

5 References

1. Zhang, M.H. and Gjorv. O.E. (1992) Penetration of cement paste into lightweight aggregate and cement paste. *Cement and Concrete Research*, Vol. 22, No. 1, pp. 47-55.
2. Zhang, M.H. and Gjorv, O.E. (1990) Microstructure of the interfacial zone between lightweight aggregate and cement paste. *Cement and Concrete Research*, Vol. 20, No. 4, pp. 610-18
3. Zhang, M.H. and Gjorv, O.E.   (1990) Pozzolanic reactivity of lightweight aggregates. *Cement and Concrete Research*, Vol. 20, No. 6, pp. 884-90.
4. Khokorin, N.K.   (1973) *The Durability of Lightweight Aggregate Concrete Structural Members*, Kuibyshev, USSR
5. Knigina, G.I. and Pinayev, A.A. (1975) Mikrokalorimetria i stabilnost phasovogo sostava Agloporita. *Stroitelny Materialy*, No. 241, pp. 29-30.

6. Fagerlund, G.. (1978) *Frost Resistance of Concrete with Porous Aggregate, Research report 2/78.* Cement and Concrete Research Institute, Sweden.
7. Swamy, R.N. and Lambert, G.N.. (1983) Mix design and properties of concrete made with PFA coarse aggregate and sand. *Int. J. Cement Composites and Lightweight Concrete*, Vol. 5, No. 4, pp. 263 -74.
8. Sarkar, S. I., Chandra. S and Berntsson, L. (1992) Interdependence of microstructure and strength of structural lightweight aggregate concrete. *Cement and Concrete Composites*, Vol. 14, No. 4, pp. 239-48.
9. Wasserman, R. and Bentur, A. (1997) Effect of lightweight fly ash aggregate microstructure on the strength of concretes. *Cement and Concrete Research*, Vol. 27, No.4, pp. 532-37.
10. Bentur, A. (1991) Microstructure, interfacial effects and micromechanics of cementitious composites, in *Advances in Cementitious Materials, Ceramics Transactions*, Vol. 16 (ed. S. Mindess). The American Ceramic Society, pp. 523-50.
11. Wasserman, R. and Bentur, A. (1996) Interfacial interactions in lightweight aggregate concretes and their influence on the concrete strength. *Cement and Concrete Composites*, Vol. 18, No. 1, pp. 67-76.

28 PROPERTIES OF INTERFACIAL TRANSITION ZONE IN GYPSUM-FREE CEMENT–ROCK COMPOSITES

Z. BAŽANTOVÁ
Czech Technical University, Civil Engineering Faculty, Prague, Czech Republic
S. MODRÝ
Czech Technical University, Klokner Institute, Prague, Czech Republic

Abstract
A contribution to the knowledge on the influence of the type of binding and rock on the character and properties of the contact zone with special respect to the aplication of the gypsum free cements. For experimental investigation of zone properties five types of rocks were used and three types of cements. Character of contact zone was investigated with the aid of microhardness measurement method, scanning electron microscope test and cement-rock bond test. From experimental results follows that the character of the contact zone depends on the type of the binding and rock used. In the case of gypsum-free cement the contact zone has expresively compact character. The microhardness data depends on the type of the uses binder and rock. Microhardness data depends on the type of used binder and rock. Microhardness of gypsum-free cements exceeds generally microhardness of the Portland cement in the contact zone so as in the cement body. Bond strength of gypsum-free cement is twofold in comparison with corresponding data of Portland cements. Bond strength of Portland cements is influenced by their composition, so as by composition of rocks. The bond strength of gypsum-free cements depends much more on the surface relief.
Keywords: gypsum-free cement, cement-rock composites, interfacial transition zone, microhardness, scanning electron microscope, bond strength

1 Introduction

In spite of the fact that there are great differences in bond strength between cement and different rocks known at 20tieth of this century aggregates in concrete was generaly considered as inert. Knowledge on some reactions between aggregates and cement led to the intensive study of interfacial transition zone. Experimentally was proven that connection between aggregates and cement in concrete influences not only physical and mechanical properties of mortars and concretes but also pronouncedly their durability [1 – 10].

The Interfacial Transition Zone in Cementitious Composites, edited by A. Katz, A. Bentur, M. Alexander and G. Arliguie. Published in 1998 by E & FN Spon, 11 New Fetter Lane, London EC4P 4EE, UK, ISBN: 0 419 24310 0

Bond strength between aggregates and cements is influenced by chemical and mineralogical character of rocks, first of all roughness of contact surfaces. The decrease of rock surface relief makes decrease of the bond strength. Bond strength is also influenced by rock porosity. Generaly, with increasing of porosity bond strength increases.

Naturally, bond strength is influenced also by the character of the cement binder. That is why an attention was intensively paid to the research of transition zone of Portland cement and for blended cements (e.g. [8]). However, transition zone between rocks and gypsum-free cements was not matter of interest until now. That was a reason for large experimental research project realized at the Czech Technical University in the last years. The aim of this project was to contribute to the knowledge about character and properties of the interfacial transition zone esp. between gypsum-free cements and some rocks.

2 Experimental

For experimental research of the structure, physical and mechanical properties of the transition zone rocks with different chemical and mineralogical character: quartz, granite, sandstone, limestone and marble. Porosity differences were also taken into account. Rocks chosen were characterized from the point of view of their petrography, chemical and physical composition. List of used rocks with their physical-mechanical properties is given in the Table 1. Besides above given properties also roughness of sawed planes of used rocks was measured with the aid of microprofilometry method.

Table 1. Physical-mechanical properties of used rocks

Property	Rocks				
	Quartz Švedlár (Slov.)	Granite Hudčice	Sandstone Hořice	Limestone Vrača (Bul.)	Marble Slivenec
Specific weight (kg.m^{-3})	2650	2767	2656	2687	2738
Volume weight (kg.m^{-3})	2616	2733	1911	2319	2692
Porosity (%)	1,27	1,23	28,05	13,7	1,68
Water absorption (%)	0,34	0,33	10,0	3,77	0,17
Compressive strength (MPa)	320,0	247,0	26,0	108,0	145,0
Bending strength (MPa)	33,9	18,4	3,3	15,1	13,9

Besides gypsum-free cement (formerly denoted as modified quick-setting cement-MRVC [4]) ordinary Portland cements and sulfate resistant Portland cements were used. Properties of cements used are given in the Table 2.

Table 2. Physical-mechanical properties of used cements

Property	Cements			
	Maloměřice	Schwanebeck (Germany)	Maloměřice	Maloměřice
	PC 400	PZ 2/30	MRVC 0	MRVC III
Specific weight (kg.m^{-3})	3 110	3 120	3 130	3 130
Specific surface area (m^2.kg^{-1})	365,4	371,3	309,0	708,6
Fineness				
Rests on sieves				
0,2 (%)	0,1	0,4	-----	0,0
0,09 (%)	2,0	6,0	-----	0,4
0,063 (%)	5,6	14,4	-----	1,2
Quantity of water and admixture (%)	31	28	22 0,4 Na$_2$CO$_3$	23 0,4 Na$_2$CO$_3$ 0,2 Kortan
Initial setting time (hrs, min)	2,55	2,40	0,25	0,15
Final setting time (hrs, min)	5,05	4,30	0,35	0,23
Soundness	Accept.	Accept.	Accept.	Accept.
Bending strength (MPa)				
1 day	1,76	2,45	3,80	4,20
3 days	3,89	3,86	5,30	5,80
7 days	5,09	4,29	5,82	8,50
28 days	7,28	6,20	8,87	10,6
Compressive strength (MPa)				
1 day	6,65	10,2	10,5	21,4
3 days	22,7	23,4	22,4	25,9
7 days	32,9	28,4	34,6	69,8
28 days	44,3	37,1	58,2	80,3

For preparation of the specimens for testing of cement-rock bond testing large blocks of rocks were sawed into prisms of sizes 40 x 40 x 80 mm. These prisms were put to steel mold of sizes 40 x 40 x 160 mm and the mold was after that filled up by cement paste of standard consistency. Cement paste was compacted by the vibration with the aid of vibrating table. Testing specimen were left for 24 hrs in molds at 90%

relative humidity and temperature 20°C. After demolding specimens were left at same conditions up to 28 days.

Bond strength between rocks and cement pastes was tested at the age of 28 days by bending test with the aid of four point test in analogy with formerly used procedures [1, 4, 9]. On rests of specimens compressive strength of cement paste was tested.

Physical property-microhardness and morphology of transition zone was tested with the aid of microhardness method and scanning electron microscope. Also pore structure of used rocks and hardened cement pastes was determined in experiments. Results are given in [4].

3 Results and their discussion

3. 1 Surface properties
Basic parameters of roughness of sawed rock surfaces are presented in the Table 3.

Table 3. Roughness of sawed surfaces of used rocks.

Parameters	Rocks				
	Quartz Švedlár (Slov.)	Granite Hudčice	Sandstone Hořice	Limestone Vrača (Bul.)	Marble Slivenec
Mean deviation of profile R_a (um)	3,93	6,00	29,43	6,10	2,86
High of unevenness of profile from 10 points R_z (um)	26,4	27,6	132,4	33,6	12,8
Highest high of unevenness of profile R_m (um)	38	36	152	42	24

Quartz has relatively smooth surface of sawed surface with some cracks. Marble had smoothest surface in comparison with other rocks. Highest roughness of sawed surface has sandstone.

As it was already stated roughness of saw surface depends on character of the rocks incl. their pore structure. This statement is supported by results of pore structurel analysis published elsewhere [4]. Sequence of rocks with respect to their surface roughness is practically identical with sequence of rocks as concerns with their pore characteristics.

3. 2 Bond strength
Bond strength results of interfacial transition zone are given in the Table 4.

From results in the Table 4 follows, that bond strength between rocks and gypsum-free cements is higher in comparison with Portland cements. Values of bond strength of gypsum-free cement, type III reaches higher values than bond strength of gypsum-free cement type 0.

Table 4. Bond strength at the age of 28 days

Cements	Bond strength (MPa)				
	Rocks				
	Quartz Švedlár (Slov.)	Granite Hudčice	Sandstone Hořice	Limestone Vrača (Bul.)	Marble Slivenec
Ordinary Portland cement PC 400	3,46	3,01	2,61	4,63	3,70
Sulfate resistant Portland cement PZ 2/30	3,32	3,13	1,84	3,42	3,30
Gypsum-free cement type 0	5,49	4,83	2,52	8,85	4,26
Gypsum-free cement type III	7,82	6,21	2,89	10,05	5,94
Ordinary Portland cement PC 400 + plasticizer	3,69	3,86	2,30	4,12	3,79
Sulfate resistant Portland cement PZ 2/30 + plasticizer	3,43	4,11	2,05	3,28	4,04

Generaly higher values of bond strength in the case of gypsum-free cements are based not only on difference in the fineness of cements but also on different reological and other properties of gypsum-free cements influenced by admixtures used for regulation of setting and hardening and reology. These special properties of gypsum-free cement pastes are manifested in different properties of transition zone, e. g. higher microhardness and lower porosity [4]. The only exception is zone with sandstone. In this case testing specimens were broken in majority in the body of sandstone and not in transition zone as it is the case of other composites. That is why it was not possible to objectify results obtained.

Standard consistency of all pastes was kept constant. Due to this measure gypsum-free cement pastes had lower water-cement ratio. According to majority of authors water-cement ratio has strong influence on strength. Therefore, it was necessary to find to what extent are bond strength values influenced by quantity of water in cement pastes.

From results given in Table 3 and obtained on specimens made of Portland cement and where water-cement ratio was reduced by plasticizer for keeping the same consistency follows that even after reduction of water-cement ratio bond strength does not reach values of bond strength of comparable gypsum-free cement, type 0.

Larger difference between bond strength values for different cements appeared only for Portland cements PC 400 and PZ Z/30 (differing in content C_3A) in the case of limestone. Under comparable conditions of producing and curing testing specimens cement PC 400 has higher bond strength for composite with limestone in comparison with bond strength with quartz. Simultaneously, the difference in bond strength for low aluminate cement PZ 2/30 and mentioned rocks negligible. This fact fits with findings published in [5] and shows the influence of C_3A content on mechanical properties of mortars produced with siliceous and calcareous aggregates.

Comparison of SiO_2 content in siliceous rocks used in the experiment with bond strength values showed fact that the higher is SiO_2 content, the higher is bond strength. It corresponds to conclusions published in [1]. In the case of tested calcareous rocks this relationship between content of CaO and bond strength is reverse. Naturally, it is necessary to bear in mind also different porosity and its influence on character of contact between rock and cement binder [4].

From comparison of bond strength values with surface characteristics of rocks used follows that bond strength of quartz, granite and marble with Portland cements depends only little on relief changes. Bond strength with sandstone is not discussed here due to character of rupture described before. In the case of limestone with surface properties close to surface properties of granite bond strength of limestone is higher in comparison with granite. This fact is valid for PC 400 and for both types of gypsum-free cement but not for sulfate resistant Portland cement PZ 2/30. Difference in bond strength between limestone and granite is negligible. This fact is apparently connected with different content of C_3A in cements used [4].

Dependence of bond strength of gypsum-free cements on relief character of rock surface is pronouncedly higher and almost linearly increases in sequence as follows: marble-quartz-granite. The highest bond strength values are on the contact of gypsum-free cements and limestone; with absolutely highest are values of gypsum-free cement, type III.

4 References

1. Alexander, M., Wardlaw, J. and Gilbert, D.J. (1968) *The Structure of Concrete and its Behaviour under Load*, (ed. A.E.Brooks and K.Newman), Cem. Concr. Assoc., London, pp. 59 – 81.

2. Barnes, B.D., Diamond, S. and Dolch, W.L. (1979) Micromorphology of the interfacial zone around aggregates in Portland cement mortar. *Journal of the American Ceramic Society*, Vol. 62, No. 1 – 2. pp. 21 – 24.

3. Barnes, B.D., Diamond, S. and Dolch, W.L. (1978) The contact zone between Portland cement paste and glass "aggregate" surfaces. *Cement and Concrete Research*, Vol. 8, No. 2. pp. 233 – 244.

4. Bažantová, Z. (1985) *Bond between cement binder and rocks*, (PhD Thesis), Faculty of Civil Engineering of the Czech Technical University, Prague.

5. Chatterji, S. and Jeffery, J.W. (1971) The nature of the bond between different types of aggregates and Portland cement. *Indian Concrete Journal*, Vol. 45, Aug. pp. 346 – 349.

6. Langton, C.A. and Roy, D.M. (1980) Morphology and microstructure of cement paste / rock interfacial regions, in 7th *Intern. Congress on the Chemistry of Cement*, Vol. III. Commun. (Suite), Editions Septima, Paris, pp. VII – 121 – 132.

7. Ljubimova, R.Ju. and Pinus, E.P. (1962) Processy kristallizacionovo strukturoobrazovanija v zone kontakta mezdu zapolnitelem i vjazuscin v cementnom betone. Kolloidnyj zurnal, Vol. 24, No. 5. pp. 578 – 587.

8. Maso, J.C. (1980) The bond between aggregates and hydrated cement paste, in 7th *Intern. Congress on the Chemistry of Cement*, Vol. I., Sub – theme VII – 1, Editions Septima, Paris, pp. VII – 1/1 – 15.

9. Struble, L., Skalny, J. and Mindess, S. (1980) A review of the cement – aggregate bond. *Cement and Concrete Research*, Vol. 10, No. 2. pp. 277 – 286.

10. Valenta, O. (1961) The significance of the aggregate – cement bond for the durability of concrete, in RILEM Intern. Symp. Durability of Concrete. Preliminary Rept., Publ. House of the Czech. Academy of Sciences, Prague, pp. 53 – 87.

29 INFLUENCE OF INITIAL MICROSTRUCTURING IN THE ITZ ON EARLY HYDRATION-STRENGTH BINDING

C. LEGRAND
Laboratoire Matériaux et Durabilité des Constructions, Toulouse, France
E. WIRQUIN
Laboratoire d'Artois Mécanique et Habitat, Béthune, France
in collaboration with l'Association Technique de l'Industrie des Liants
Hydrauliques

Abstract
This paper concerns observations and interpretations of changes in mortar strength at the very beginning of its lifetime (0 to 24 hours) in relation to advance of hydration, taking a special interest in the mineralogical nature of the aggregates. During the first steps of hydration (dormant period), the effects of the consolidation of the bonds are not the same and it seems that this period corresponds to the real birth of the future I.T.Z..Indeed, during and just after the setting, it appears that the I.T.Z. has inherited their properties from the first structuring of the initial bonds. The use of superplasticizers has a strong influence on these bonds and tends to amplify these phenomena.
Keywords : Hydration, I.T.Z., limestone and siliceous aggregates, strength, superplasticizer.

1 Introduction.

Legrand and al [1] have shown the role played by the initial geometrical arrangement of the cement grains (microstructuring) in binding efficiency between the quantity of formed hydrates and acquisition of strength of a mortar during the so-called dormant period, the setting and the first 24 hours. Using results published by Mutin and al [2], they represented a structural model allowing the interpretation they ascribe to the observed phenomena to be clearly demonstrated. This model is based on the flocculating structure of cement paste without additive [3] where intergranular links by Coulomb and Van der Waals attractions are demonstrated.

The interest of a study on mortar rather than on pure paste resides in the fact that it allows to take into account the strengthening of bindings between cement grains and aggregates, from the smallest to the largest, which, as everyone knows, plays an important role [4]. This role can be made obvious if we change the mineralogical nature of aggregates.

The Interfacial Transition Zone in Cementitious Composites, edited by A. Katz, A. Bentur,
M. Alexander and G. Arliguie. Published in 1998 by E & FN Spon, 11 New Fetter Lane,
London EC4P 4EE, UK, ISBN: 0 419 24310 0

2 Experimental process.

Tests have been carried out on mortars made with the same cement, dosed to 500 kg/m³ and with a water / cement ratio equal to 0.45. A superplasticizer dosed to 1% has or has not been added to these mortars and water content was kept constant. The granular part was constituted of limestone or siliceous materials. The granulometric reference curve has been reconstituted in both cases, after washing aggregates carefully so as to eliminate the fillers always more or less present on mineral surfaces of this nature. The experimental procedure has already been described [5] and a brief summary is given hereafter.

During the dormant period and the beginning of the setting, the mechanical strength is represented by the shear stress at failure τ_f, as a function of time, by means of a viscometer transformed for the opportunity into vane-test. As soon as the mortar can be removed from the mould, prismatic test pieces have been made and submitted to a normalized test giving the compressive strength fc. This procedure therefore allows to follow the mechanical strength evolution of mortars τ_f (t), then fc (t) since the end of the mixing until a limit fixed to 24 hours.

Semi-adiabatic calorimetric tests have been carried out simultaneously on the same mortars, that have allowed access to the specific hydration heat which is supposed to represent, in first approximation, the global quantity of hydrates generated by 1 gram of cement. The evolution of temperatures being different in mortars and in the calorimeter, corrections have been made to work with equivalent time so as to take account differences of maturity [6]. When this operation is realized, we can obtain relationships τ_f (q) and fc (q) that allow to appreciate the increase of mechanical strengths according to the increase of the quantity of formed hydrates, independently of the time required.

3 Study of the period between the end of the mixing and the setting.

This period concerns the time when the mortar is not removable and when its mechanical strength is appreciated with the failure shear stress τ_f .

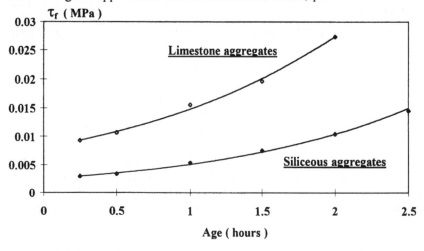

Fig. 1. Shear strength τ_f according to the time for the plain limestone and siliceous aggregates mortars.

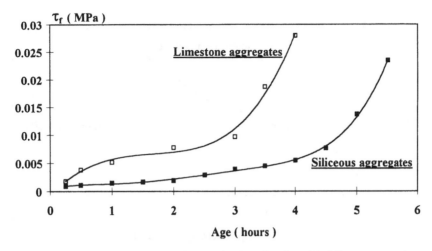

Fig. 2. Shear strength τ_f according to the time for the added limestone and siliceous aggregates mortars.

Figures 1 and 2 give curves τ_f (t) for siliceous and limestone aggregates mortars whether they are added or not.

It is similar for the heat development q (t) (figures 3 and 4).

Concerning mechanical strengths, we undoubtedly observe a best systematic performance of the limestone aggregates mortar, whether it is added or not.

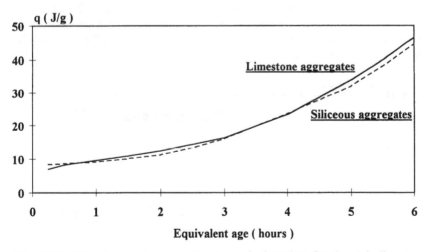

Fig. 3. Heat development q according to equivalent time for the plain limestone and siliceous aggregates mortars.

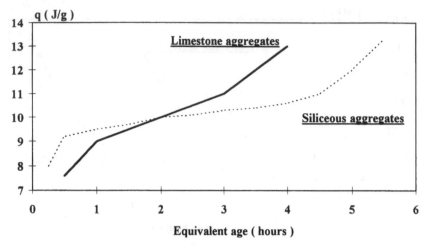

Fig. 4. Heat development q according to equivalent time for the added limestone and siliceous aggregates mortars.

On the other hand, the mineralogical difference between aggregates does not seem to have any influence on the heat development for the plain mortar. In the case of the added mortar, if evolutions are related until 2 hours approximately, the limestone aggregates mortar seems to produce more hydrates than the other one. It would appear thus, that after a certain time, there is an interaction between the admixture and the surface of aggregates and that this phenomenon interferes on the rate of hydration of the cement. The molecules of sulphonate naphtalene are more easily adsorbed by limestone sites than by siliceous sites. We could then think that the limestone aggregates mortar, having less admixture adsorbed by the cement grains than the other one, would be less subject to the delay of hydration.

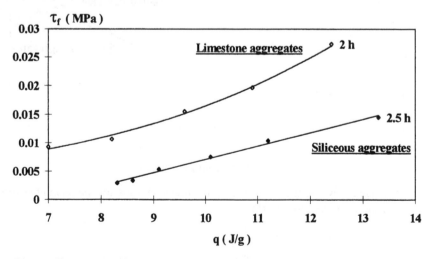

Fig. 5. Shear strength τ_f according to the heat development q for the plain limestone and siliceous aggregates mortars.

Now, if we analyze the curve τ_f (q) for plain mortars (figure 5), we observe that for quantity of equal formed hydrates, the limestone aggregates mortar is always more effective than that with siliceous aggregates. We can thus notice that the initial strength is always higher for the limestone aggregates mortar which at the beginning, is stiffer although the water content is the same. This cannot be attributed to a parasitic presence of limestone fillers considering the meticulous washing done. Measures of water absorption on the limestone aggregates mortar have not given any significant results; furthermore, we have checked that there was no important wearing-away during the mixing. We can only attribute this difference of workability to the influence of the mineralogical aggregates nature already made obvious by Maso [7], then by Legrand [8] and we could then think of a deficit in water of the cement paste of the limestone aggregates mortar compared to the siliceous aggregates mortar.

However, the paste-aggregate two-phase model surrounded by water [9] seems to be insufficient now to suitably describe the influence of the aggregates. We could rather think of a very early role played by the interaction between cement grains and aggregates. In the short run, initial bindings and the structuring would be different according to nature of the solid phase in presence. This phenomenon could be the cause of the difference of workability and the birth of the future Interfacial Transition Zones (I.T.Z.) whose microstructure would depend partly on the initial state, which depends on the nature of aggregates.

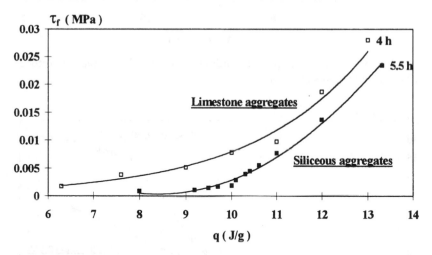

Fig. 6. Shear strength τ_f according to the heat development q for the added limestone and siliceous aggregates mortars.

Concerning the added mortar (figure 6), despite the dispersion of grains and the practically null initial strength resulting, we find a similar influence of the nature of aggregates. Thus, it would seem that, from the end of the mixing, interfacial zones between aggregates and the cement paste play an important role in the evolution of strength according to the quantity of formed hydrates. Limestone aggregates would present zones that would be more favorable to the development of strengths than siliceous aggregates, with mortar added or not, even if we are not yet able to know if the incorporation of the additive modifies the structure of the paste in these zones.

4 Study of the period between the end of the setting and 24 hours.

The mortar is now removable and we can have access to its compressive strength fc (figures 7 and 8). We note that the phenomena observed during the previous period continue : limestone aggregates make the mortar more effective at equal age and all the more effective as it will have been added.

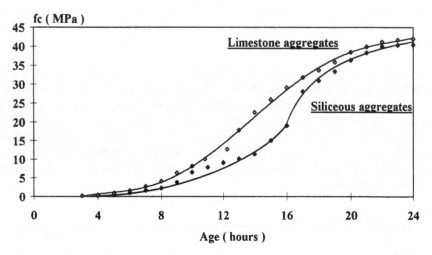

Fig. 7. Compressive strength fc according to the time for the plain limestone and siliceous aggregates mortars.

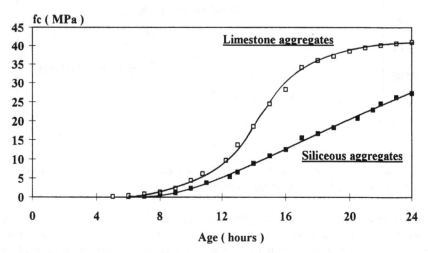

Fig. 8. Compressive strength fc according to the time for the added limestone and siliceous aggregates mortars.

The specific hydration heat (figure 9 and 10) becomes practically independent from the mineralogical aggregates nature if there is no admixture whereas it continues to differentiate in the presence of sulphonate naphtalene.

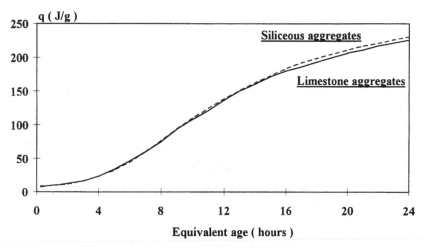

Fig. 9. Heat development q according to equivalent time (until 24 hours) for the plain limestone and siliceous aggregates mortars.

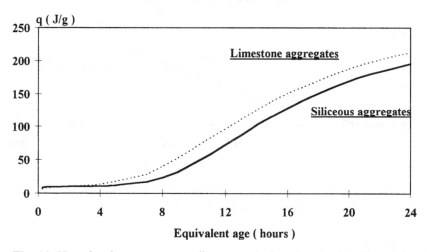

Fig. 10. Heat development q according to equivalent time (until 24 hours) for the added limestone and siliceous aggregates mortars.

Curves fc (q) (figures 11 and 12) show that the I.T.Z. in young concrete have "inherited" interfacial zone properties divulged to the fresh state. Moreover, it seems that the use of an admixture increases phenomena, which would be confirmed by strengths to 28 days. The gain of strength to 28 days is 7 % for the plain mortar and 15 % for the added mortar. It is not forbidden to think then that the additive modifies the structure of the paste in the I.T.Z., which a priori seems logical.

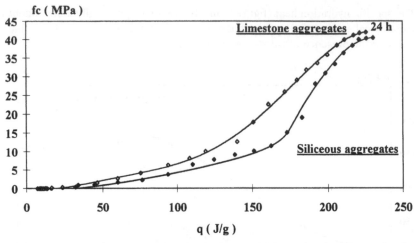

Fig. 11. Compressive strength fc according to the heat development q for the plain limestone and siliceous aggregates mortars.

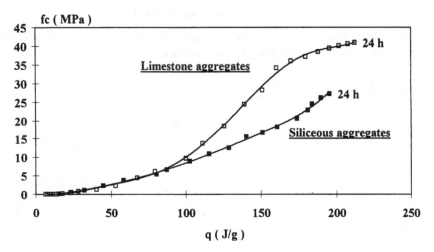

Fig. 12. Compressive strength fc according to the heat development q for the added limestone and siliceous aggregates mortars.

5 Conclusion.

It seems that I.T.Z. already exist between the cement paste and aggregates at the end of the mixing. They depend on the nature of the latter; they also depend on the presence of superplasticizers and they affect mechanical characteristics. The I.T.Z. in the young concrete inherit from the properties of interfacial zones existing at fresh state, the use of an admixture appearing amplify the phenomena.

6 References.

1. Legrand C., Wirquin E. (1994). "Study of the strengths of very young concrete as a function of the amount of formed hydrates - Influence of superplasticizers". *Materials and Structures*, Volume 27, N° 166, p. 106-109.
2. Mutin J.C., Nonat A., Eddajibi A.H. (1989). "Hydratation et prise", in *Connaissance générale du béton, Ecole Nationale des Ponts et Chaussées*, 39 p..
3. Powers T.C. (1968). "The properties of fresh concrete", John Wiley, New York.
4. Maso J.C. (1967). "La nature minéralogique des agrégats, facteurs essentiels de la résistance des bétons à la rupture et à l'action du gel", *Thèse d'Etat*, Toulouse.
5. Legrand C., Wirquin E. (1991). "First developments of strength in a microconcrete", *Hydration and Setting of Cements, Proceedings of the International RILEM Workshop*, Dijon, p. 299-306.
6. Kouakou A., Legrand C., Wirquin E. (1996). "Mesure de l'énergie d'activation apparente des ciments dans les mortiers à l'aide du calorimètre semi-adiabatique de Langavant", *Materials and Structures*, Volume 29, N° 191, p. 444-447.
7. Maso J.C. (1982). "La liaison pâte-granulats", *Le Béton Hydraulique, Presses de l'Ecole Nationale des Ponts et Chaussées*, Paris, p. 247-259.
8. Legrand C. (1971). "Etude statistique de l'existence de facteurs liés à la nature des granulats intervenant sur la viscosité des mortiers frais", *Compte rendu à l'Académie des Sciences*, Série A, p. 834-836.
9. Barrioulet M., Legrand C. (1977). "Influence de la pâte interstitielle sur l'aptitude à l'écoulement du béton frais. Rôle joué par l'eau retenue par les granulats", *Materials and Structures*, Volume 11, N° 60, p. 365-373.

30 THE INFLUENCE OF TIME-DEPENDENT CHANGES IN ITZ ON STIFFNESS OF CONCRETES MADE WITH TWO AGGREGATE TYPES

M.G. ALEXANDER
Department of Civil Engineering, University of Cape Town, South Africa
K. SCRIVENER
Lafarge, Laboratoire Centrale de Recherche, France

Abstract
The paper presents results on the effects of time-dependent changes in the ITZ on mechanical properties of concrete. Concretes were prepared with OPC at two w/c ratios (0,43 and 0,61). Two different crushed aggregates were used, derived from andesite and dolomite rock, comprising coarse and fine fractions. The compressive strength and elastic modulus of the concretes were measured up to 90 days after casting. At the various ages of testing, samples were cut from companion specimens, and subjected to a preparation regime before being examined under a SEM for ITZ properties of porosity, percentage anhydrous material, and quantity of CH. The results, although only preliminary, suggest that time-dependent changes in stiffness of concretes made with different aggregates may be explained in terms of development of the characteristic properties of the ITZ.
Keywords: Interfacial transition zone, aggregates, stiffness, microstructure.

1 Introduction

The nature of the so-called interfacial transition zone (ITZ) in cementitious composites has been the subject of study and debate for at least 4 decades. More recently, there has been an international conference on the subject [1], and an increasing amount of published information. Much of the research data relates to microstructural studies of the morphological nature of the ITZ [2], and methods of characterising its essential physical and chemical properties. Other studies have shown the link between the properties of the ITZ in cement-fibre composites and the mechanical behaviour of the material [3]. However, there is very little information demonstrating direct links between ITZ properties and mechanical behaviour of normal portland cement concretes. The objective of this paper is to attempt to explore such links.

The Interfacial Transition Zone in Cementitious Composites, edited by A. Katz, A. Bentur, M. Alexander and G. Arliguie. Published in 1998 by E & FN Spon, 11 New Fetter Lane, London EC4P 4EE, UK, ISBN: 0 419 24310 0

Previous work has been done on a broad range of South African crushed rock aggregates, in order to characterise the elastic modulus - compressive strength relationships [4,5]. Each aggregate type or combination was found to have a 'unique' E-f_{cu} relationship, which was sensibly linear for cube strengths between 20 MPa and 60 MPa. A series of such relationships is reproduced from reference [5] in Figure 1.

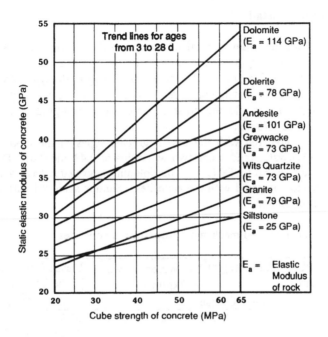

Figure 1. Elastic modulus-strength relationships for concretes with different aggregate types (from ref. [5])

While the E-f_{cu} relations are useful from a practical engineering design point of view, it would be helpful to explore mechanisms for the different behaviours. For example, concretes with aggregates having approximately equal rock elastic moduli may exhibit different elastic modulus relationships – see 'Dolerite' and 'Granite' in Figure 1. Furthermore, certain aggregate types produce concretes with slopes of the E-f_{cu} relationships that are essentially constant with time (for example the andesite aggregate in [5]). On the other hand, dolomite aggregate concretes have E-f_{cu} relationships at early ages with steeper slopes when compared with later-age relationships. In most cases, concretes stiffen with age to an extent that cannot be accounted for purely in terms of increased strength. These observations raise questions concerning influencing mechanisms, and in particular the influences that the ITZ and its development with time may have on the mechanical properties of the concrete.

Quantitative studies of the microstructure in the ITZ [6, 7, 8] indicate that the predominant factor is the packing of the cement particles against the aggregate. This leads to a deficit of anhydrous material in this region and an increase in porosity. The increased porosity and mobility of calcium ions allow the growth of more calcium hydroxide (CH) in this region.

In order to investigate these issues, an experimental programme was designed as a collaborative effort between the two authors at their respective institutions at that time, i.e. the University of the Witwatersrand, Johannesburg and Imperial College, London. The primary object of the work was to explore the relationships between the engineering macro-properties of concretes and the micro-properties of the ITZ.

2   Experimental details

The cement used was an ordinary portland cement (ASTM Type I). Coarse and fine aggregates were of two types derived from crushing of fresh rock: a strong fine-grained andesite, and a fine-grained dolomite. Both aggregate types are commonly used in concrete in the Johannesburg area. The 19 mm coarse aggregates were nominally of single size, while the crusher sands were well-graded materials with maximum particle size 4,75 mm.

Concretes were prepared with the two aggregate types in the laboratory at the University of the Witwatersrand, using two w/c ratios of 0,43 and 0,61. Details of the mix proportions are shown in Table 1. The dolomite aggregates have a substantially lower water demand compared with the andesites. Slumps achieved with these mixes were of the order of 50 mm.

Table 1. Mix proportions and mechanical properties of concretes

Aggregate Type	OPC (kg/m^3)	Aggr. (kg/m^3)		Water (kg/m^3)	w/c	Elastic Modulus (GPa) [Cube Strength (MPa)]		
		Fine	Coarse			2 d	28 d	90 d
Dolomite	272	908	1210	165	0,61	33,7 [17,2]	47,0 [34,0]	51,0 [41,5]
	400	711	1280	170	0,43	42,7 [27,7]	54,3 [56,5]	58,0 [62,6]
Andesite	314	916	1130	190	0,61	27,7 [14,9]	38,7 [36,7]	40,7 [44,3]
	458	685	1200	195	0,43	36,0 [27,7]	44,0 [65,7]	44,7 [72,4]

100 mm cubes and 100 x 100 x 200 mm prisms were cast, the moulds stripped at 24 hours, and the specimens water-cured at 23°C. Sufficient cubes were cast to permit 3 cubes to be compression tested at 2, 28, and 90 d after casting. At the same ages, three prisms were tested for elastic modulus (E) in compression. Special care was taken for the 2 d test to limit the compressive stress to less than 10% of the 2 d strength to prevent damage to the prisms. The same set of 3 prisms was used for all the E tests. At 28 d and 90 d, the compressive stress used was approximately 25% of the strength at that age.

Samples for SEM analysis were prepared as follows. Additional prisms were cast from each mix, and two 20 x 50 x 100 mm slices were cut at mid-height using a fine-bladed diamond saw, at ages of 2, 28, and 90 d. The orientation of the slices as cut is shown in Figure 2. One slice from each mix was oven-dried for 7 d in a ventilated

50°C oven (15-25% R.H.), then sealed in a plastic bag and dispatched to Imperial College, where they were further prepared for SEM observation. The second slice was further cut at mid length to produce 4 wafers 20 x 50 x 5 mm thick, which were freeze-dried by immersion in a solution at -80°C till frozen, followed by 1 week of drying. These wafers were likewise sealed after drying and sent to Imperial College.

The cut samples were impregnated with resin and polished to 0,25 μm, coated with carbon and examined in the SEM (JEOL 35CF) with KONTRON – IPS image analysis system. For each sample 25 images were selected at random around the aggregate interfaces, and were analysed at a magnification of 600x. For each image the area fractions of anhydrous, calcium hydroxide and porosity were measured in 30 bands, 3,4 μm thick radiating out from the aggregate surface.

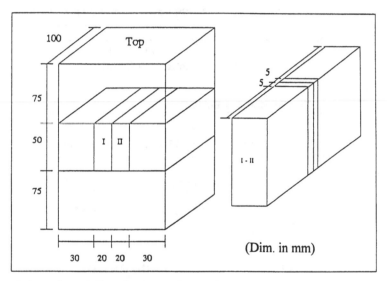

Figure 2. Orientation of slices for SEM observation

3 Results and Discussion

3.1 Mechanical properties of concretes
Table 1 shows elastic modulus and compressive strength results. The stress-strain plots from which E values were determined all showed good linearity, even at early ages, allowing a value for E to be confidently defined for each test. In Figure 3, E is plotted against cube strength f_{cu}. Linear regression relationships are shown for those points falling sensibly on a straight line. For the andesite concretes, only the 2 d, w/c=0,61 result is not co-linear with the other results; for the dolomite concretes, both 2 d results fall below the regression line, implying that stiffness of dolomite concrete develops at a slower rate at early ages than andesite concrete, on an equal compressive strength basis. Figures 3 (a) and (b) also clearly demonstrate that dolomite concretes attain substantially higher E_c values than andesite concretes at equal compressive strength.

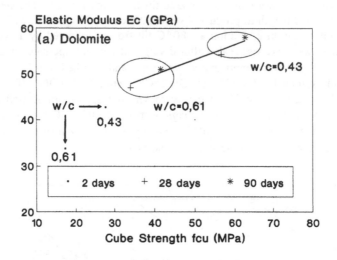

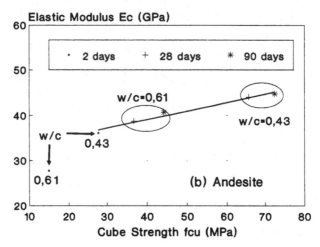

Figure 3. Elastic modulus – cube strength relationships for (a) dolomite and (b) andesite concrete

3.2 Results of microstructural studies

Previous work [9] has shown that 25 images is a low number to get results with low variability (more couldn't be done due to time constraints). Nevertheless the image analysis gave clear variability in the anhydrous measurements, particularly at >20 µm from the interface. Thus, the discussion in this paper will focus on the anhydrous content (% ANH) in the 20-40 µm zone. (With the low number of images analysed, the parameters of porosity and CH were judged to give inconsistent results; it is these same parameters that are most susceptible to the techniques used for sample preparation).

Figures 4(a) and (b) show the variation with distance from dolomite and andesite aggregate particles of unhydrated cement fraction (anhydrous) respectively, for the ages of test, at w/c = 0,61. It is clear that at 2 days, the dolomite samples have higher

ANH than the andesite samples in the 20 - 40 μm zone. Figures 5(a) and (b) give similar data for w/c = 0,43. The same trend at 2 d of higher ANH in the 20 – 40 μm zone for the dolomite samples is also seen here.

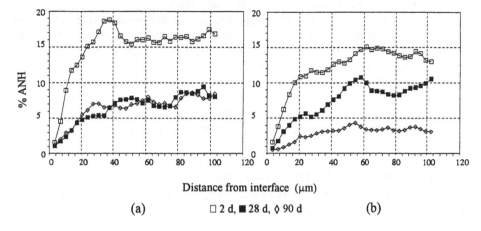

Distance from interface (μm)

(a)   □ 2 d, ■ 28 d, ◊ 90 d   (b)

Figure 4. Variation of ANH with distance from aggregate particles of (a) dolomite and (b) andesite concrete, w/c = 0,61

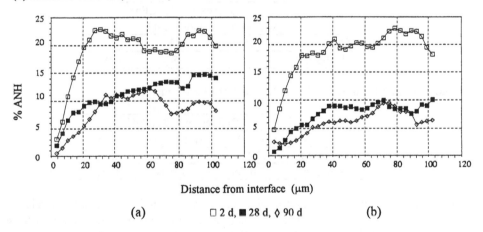

Distance from interface (μm)

(a)   □ 2 d, ■ 28 d, ◊ 90 d   (b)

Figure 5. Variation of ANH with distance from aggregate particles of (a) dolomite and (b) andesite concrete, w/c = 0,43

At later ages (28 d and 90 d) the general trend is that substantial reductions in ANH have occurred, due to continuing hydration. It is clear from Figures 4 and 5 that, for mature concretes, the dolomite samples have considerably higher quantities of ANH in the 20 – 40 μm zone in comparison with the andesite sample. The possible implications of this will be further explored below.

3.3   General discussion
Considering the effects of the ITZ on concrete stiffness (E), two important effects will

influence this stiffness:

- continuing hydration in the ITZ (and the bulk paste), resulting in pore-filling by hydration products; this effect will be more pronounced in the early stages between 2 d and 28 d, when the rate of hydration is fairly high.
- the amount of ANH in the ITZ, since the unhydrated cement represents a well-bonded and relatively stiff micro-aggregate, thus enhancing stiffness. This effect will be more important in mature concretes where rate of hydration is very slow and pore-filling is essentially accomplished.

These two effects are somewhat competitive in influencing stiffness, in that further hydration will result in pore-filling, thus increasing stiffness, but will also reduce the amount of ANH which acts as a stiff micro-aggregate, thus reducing stiffness.

With the above framework in mind, it is appropriate to consider the results for the time period of 2 - 28 d, and at 90 d.

Effects in the period 2 – 28 d (20 – 40 μm zone)
Effects in this period are best shown by examining the increase of E (Table 2) in relation to the reduction of ANH (Table 3).

Table 2. Increase in E from 2 – 28 d, for both concretes and both w/c ratios

	Increase in E (GPa)		
	w/c = 0,61	w/c = 0,43	Mean
Dolomite	15	11	13
Andesite	11	8	9

Table 3. Reduction in % ANH from 2 – 28 d, for both concretes and both w/c ratios

	Reduction in % ANH		
	w/c = 0,61	w/c = 0,43	Mean
Dolomite	11,5 %	12,5 %	12 %
Andesite	5,0 %	11,0 %	8 %

The dolomite samples have experienced a greater increase in E, with a correspondingly greater reduction in ANH, over this period. In contrast, the andesite concretes have experienced lower E increases and ANH reductions. Using the mechanism of continuing hydration and pore-filling, this would seem to indicate that at least a portion of the greater E increases of the dolomite concrete might be explained by a low early age (2 d) development of stiffness, which improves at later ages. In addition, the dolomite concretes appear to be much more sensitive to changes in ANH than the andesite concretes.

Effects at 90 d (20-40 μm zone)
For both w/c ratios, mature dolomite concretes are substantially stiffer than andesite concretes. These differences amount to between 10 to 14 GPa, and are much larger than would be accounted for on the basis of the small differences in aggregate elastic moduli ($E_{andesite}$ = 101 GPa, $E_{dolomite}$ = 114 GPa). A calculation using a two-phase Series Model shows that the difference in E between the two concretes at this mature

age should be of the order of 2 GPa at most.

Examination of the mean ANH in the 20 – 40 µm zone gives the following (Table 4).

Table 4. Mean % ANH in the 20 – 40 µm zone, at 90 d, for both concretes and both w/c ratios

	Mean % ANH	
	w/c = 0,61	w/c = 0,43
Dolomite	6 %	7 %
Andesite	3 %	5 %

It is clear that dolomite concretes have much greater amounts of ANH acting as stiff inclusions in the ITZ. Whether these differences are sufficient to account for the large differences in E is not clear, and is an area that could be further explored.

4   Conclusions

A series of concretes with w/c = 0,61 and 0,43, made with two different aggregates – crushed dolomite and andesite – were studied for their ITZ microstructure and $E - f_{cu}$ relationships as a function of age, between 2 d and 90 d after casting. The results are tentative, and are used to illustrate techniques and effects that might fruitfully be explored in attempting to relate microstructure to engineering properties of concrete. Considerable extra work is required to confidently define the governing mechanisms and relationships. Tentative conclusions are:

1. At early age (2 d), dolomite concrete samples show much higher ANH in the 20 – 40 µm zone than andesite concretes, for both w/c ratios. At later ages (28 d and 90 d), substantial reductions in ANH occur, but with dolomite concretes having higher residual ANH than andesite concretes.
2. The competing effects of continuing hydration at relatively early ages and the remaining amount of ANH in the ITZ, will influence the development of stiffness.
3. In the period 2 d to 28 d, there is a broad correlation between increase in E and reduction in ANH, with dolomite concretes showing the larger changes. This is consistent with the behaviour of dolomite concretes which tend to stiffen more at later ages than andesite concretes.
4. At 90 d, the more important effect seems to be that of residual ANH in the ITZ, since unhydrated cement represents a well-bonded, stiff micro-aggregate. The dolomite concretes contain up to twice as much ANH in this zone at later ages, which may help to explain their considerably greater stiffness, despite the fact that the two rock types have similar elastic moduli.

In summary, dolomite concretes show larger changes in ITZ with time in comparison with andesite concretes. These observations are consistent with the age-dependent mechanical properties of the concretes, where dolomite concretes generally show lower early development of stiffness, but also improved stiffness at later ages in comparison with andesite concretes. The ITZ in dolomite concretes appears to be more sensitive to ageing processes, whereas the important properties of the ITZ and andesite concretes are developed earlier, and more steadily with time.

5   Acknowledgements

The authors are indebted to the two students, A Stamatiou at Wits University, and S Jensen, Imperial College, who carried out the experimental work. Financial support for the South African component of the work came from the S.A. Cement Industry, via the Portland Cement Institute, and from the Foundation for Research Development.

6   References

1. *Interfaces in Cementitious Composites*, (ed. J.C. Maso), E & FN Spon, London, 1992.
2. Scrivener, K.L., Bentur, A. and Pratt, P.L. (1988) Quantitative characterisation of the transition zone in high strength concretes, *Advances in Cement Research*, Vol. 1, pp. 230-9.
3. Bentur, A. (1990) Microstructure, interfacial effects, and micromechanics of cementitious composites, *Advances in Cementitious Materials*, (ed. S. Mindess), The American Ceramic Society, pp. 523-49.
4. Alexander, M.G. (1991) An experimental critique of the BS8110 method of estimating concrete elastic modulus, *Magazine of Concrete Research*, Vol. 43, No. 157, pp. 291 -304.
5. Alexander, M.G. (1993) Two experimental techniques for studying the effects of the interfacial zone between cement paste and rock, *Cement and Concrete Research*, Vol.13, No.3, pp. 567-75.
6. Scrivener, K.L. and Gartner, E.M. (1988) Microstructural gradients in cement paste around aggregate particles, *Bonding in Cementitious Composites*, (eds. S. Mindess and S.P.Shah), Materials Research Society, Vol. 114, pp. 77-86.
7. Scrivener, K.L., Crumbie, A.K. and Pratt, P.L. (1988) A study of the interfacial region between cement paste and aggregate in concrete, *Bonding in Cementitious Composites*, (eds. S. Mindess and S.P.Shah), Materials Research Society, Vol. 114, pp. 87-8.
8. Scrivener, K.L. and Pratt, P.L. (1992) *Characterisation of interfacial microstructure*, Interfacial Transition Zone in Concrete, (ed. J.C. Maso), RILEM Report 11, E & FN Spon, London, pp. 3-17.
9. Crumbie, A.K., PhD Thesis, University of London, 1994.

31 INTERFACIAL FRACTURE BETWEEN CONCRETE AND ROCK UNDER IMPACT LOADING

S. MINDESS
Department of Civil Engineering, University of British Columbia, Vancouver, BC, Canada
K.-A. RIEDER
On leave from the Institute of Applied and Technical Physics, Technical University of Vienna, Vienna, Austria

Abstract
An instrumented drop impact machine was used to study the interfacial fracture between concrete and rock using both limestone and granite. Two different types of "interfacial surface" were investigated. One was both flat and smooth; the other was obtained by splitting a limestone or granite block through impact loading, thereby producing a rough surface. Concrete was then cast on top of each type of surface. These blocks were subjected to splitting tension along the interfacial plane. The results in terms of the fracture energy and the splitting tensile strength are discussed; it was found that the bond strength is higher under impact loading, and that the bond between concrete and granite was stronger than that between concrete and limestone.
Keywords: Impact, concrete, rock, interface, fracture.

1 Introduction

One of the objectives of interface fracture mechanics is to define material properties that characterise the fracture resistance of interfaces. In general, in early work on this topic, strength was the main parameter considered. However, in more recent research, the emphasis is shifting towards parameters such as fracture energy and fracture toughness. In the last few years, a number of studies on this topic have been carried out by Rice [1], Suo [2], Hutchinson [3], Stankowski et al [4, 5], Tschegg et al [6, 7], Hassanzadeh [8], and Mindess et al [9, 10], but all of these studies were carried out under quasi-static loading conditions. Although their results led to a better understanding of the fracture processes of interfaces, there are few, if any, studies on the dynamic fracture of bimaterial interfaces. One of the aims of the current investigation was to find out whether the dynamic fracture behaviour of interfacial fracture between concrete and could be studied using an instrumented drop weight impact machine. The results of these tests in terms of fracture energy and splitting tensile strength are discussed and compared with the results of static tests.

The Interfacial Transition Zone in Cementitious Composites, edited by A. Katz, A. Bentur, M. Alexander and G. Arliguie. Published in 1998 by E & FN Spon, 11 New Fetter Lane, London EC4P 4EE, UK, ISBN: 0 419 24310 0

2 Experimental procedure

2.1 Impact tests

An instrumented drop-weight impact machine was used to generate the impact loads. This machine is described in detail elsewhere [11, 12, 13]. Briefly, it is capable of dropping a mass of 543 kg from heights up to 2.4 m onto the specimen, giving a kinetic energy of up to 11,590 Joules. The impact loads were applied as line loads along the centre lines of the top and bottom faces of the cubes. This induces tensile stresses perpendicular to the direction of loading ("splitting tension"). Attached to the bottom of the falling mass is the striking tup, which is instrumented with four electric resistance strain gauges, connected together to form a Wheatstone bridge circuit. The specimen rests on a linear support, which is also instrumented. The signals from the tup and the support were recorded at 4 μs intervals, using a high speed PC-based data acquisition system.

After signal processing, the load as a function of time was obtained. The kinetic energy transferred from the falling hammer to the specimen was simply considered as the fracture energy in the impact tests, neglecting the kinetic energies of the specimen halves after the samples were split. The work W_{ha} done by the applied load during the entire impact event can be calculated using the impulse-momentum relationship of the hammer [11]:

$$W_{ha} = \frac{1}{2} \cdot M_h \cdot \left[\left(\frac{1}{M_h} \cdot \int p(t) \cdot dt + \sqrt{2 \cdot g_c \cdot h} \right)^2 - 2 \cdot g_c \cdot h \right] \tag{1}$$

where M_h is the mass of the hammer, h is the drop height, g_c is the corrected gravitational acceleration, and p(t) is the time dependent load of the hammer during the impact event. A correction factor has to be applied to obtain the corrected gravitational acceleration, to account for frictional effects between the hammer and the guiding columns and the air resistance (g_c=0.91g, where g is the acceleration due to gravity).

2.2 Static tests

The static tests were carried out using the same shapes for the support and the tup loading device as were used for the impact tests. The Brazil or splitting test was first used by Carneiro [14] and seems to provide the most reliable data for tensile strength of concrete [15, 16]. The splitting procedure was performed using a rigid testing machine with a maximum load capacity of 150 kN operating in a displacement controlled mode with a crosshead speed of 0.5 mm/min. The value of the splitting tensile strength f_t was calculated as

$$f_t = \frac{2 \cdot P_t}{\pi \cdot a^2} \tag{2}$$

where P_t is the maximum load and "a" is the length of the side of the cube. Equation 2 was also used to calculate the dynamic splitting tensile strength, although it was originally derived for static loading conditions only. No corrections to eqn 2 were made to take the dynamic effects into account. This will be a topic for further investigations. It should also be noted that the splitting tensile strength obtained from this kind of test strongly depends on the size of the specimen [17]. Due to this size effect, two different specimen sizes were tested.

2.3 Specimens

Two specimen sizes were tested, 100 and 200 mm cubes, with each specimen composed of one-half concrete and one-half limestone or granite. The dimension of the granite or the limestone blocks were (length, width and height) 100 by 50 by 100 mm and 200 by 100 by 200 mm. The cubes were subjected to splitting tension along the interfacial plane using a 543 kg impact hammer raised to a predetermined height. Two different types of "interface surface" were investigated. The first surface was formed by cutting with a diamond saw and was both flat and smooth. The second surface was obtained by splitting a limestone or granite cube with a side length of 100 mm or 200 mm through impact, thereby producing a rough surface. Concrete was then cast on top of each type of surface. The casting direction was the same as the impact direction. Normal strength concrete (NSC) with proportions, by weight, of cement: sand: gravel: water = 1: 2: 2: 0.45 was used for this study. The maximum aggregate size was 10 mm. Companion cylinders (100 x 200 mm) were cast in plastic moulds to measure the compressive strength. All specimens, cubes and cylinders, were cured in lime-saturated water for 28 days and then tested. The compressive strength of the granite was 142 MPa, of the limestone 45 MPa, and of the NSC 42 MPa. The elastic moduli were 45 GPa, 41 GPa, and 32 GPa, respectively.

3 Results and Discussion

3.1 Static tests

Results for the Brazil test under static loading conditions are presented in Table 1. Despite the fact that the splitting strengths of granite and limestone were much lower than those reported in reference [8], the trend remains the same: Granite has the highest splitting strength, and limestone the lowest one amongst the materials included in this investigation.

Table 1. Tensile splitting strength f_t of interfaces under quasi-static and dynamic loading conditions (a = side of the cube in mm).

Type of interface	static		Drop height: 0.225 m		Drop height: 0.5 m	
	a = 100 f_t (MPa)	a = 200 f_t (MPa)	a = 100 f_t (MPa)	a = 200 f_t (MPa)	a = 100 f_t (MPa)	a = 200 f_t (MPa)
Granite – Granite	8.51	6.49	14.37	5.25	18.80	8.25
Granite - NSC (smooth)	0.94	1.05	2.17	1.83	-	-
Granite - NSC (rough)	1.54	1.75	5.52	3.07	6.92	3.92
Limestone – Limestone	2.28	2.32	7.23	3.28	6.76	5.73
Limestone - NSC (smooth)	0.81	0.78	3.58	1.63	-	-
Limestone - NSC (rough)	1.18	1.43	4.28	2.35	4.23	2.59
NSC – NSC	3.10	2.49	10.01	3.38	13.09	4.99

The different types of interfaces investigated in this study have a significant influence on the splitting strength. The value drops about 30 to 45 %, when concrete is cast on top of a smooth surface instead of on top of a rough one. Furthermore, the results show that the bonding in terms of the splitting tensile strength between granite and NSC is better than that between limestone and NSC. This was also true for the impact tests as shown below.

3.2 Impact tests

Two different impact heights - 22.5 cm and 50 cm - were chosen to investigate the influence of the loading rate on the fracture behaviour of the interfaces: This corresponds to impact velocities of 2 m/s, and 3 m/s, respectively. In addition, two specimen sizes were tested, 100 mm cubes and 200 mm cubes. The splitting tensile strength values are shown in Table 1, and the impact energies are presented in Table 2.

Table 2. Impact energy W_{ha} of interfaces under dynamic loading conditions (a = side of the cube in mm).

Type of interface	Drop height: 0.225 m		Drop height: 0.5 m	
	a = 100 W_{ha} (Nm)	a = 200 W_{ha} (Nm)	a = 100 W_{ha} (Nm)	a = 200 W_{ha} (Nm)
Granite – Granite	99	475	139	726
Granite - NSC (smooth)	29	225	-	-
Granite - NSC (rough)	47	229	66	296
Limestone – Limestone	54	225	61	285
Limestone - NSC (smooth)	38	158	-	-
Limestone - NSC (rough)	44	210	46	246
NSC – NSC	57	334	73	428

Note: the <u>specific</u> fracture energies are given in Fig. 1.

It can be clearly seen from Table 1 that the dynamic splitting strength decreases dramatically with increasing specimen size. Therefore only results obtained with the same specimen size can really be compared. It can also be seen that the trend for the dynamic tensile splitting strength of interfaces is the same for 100 mm cubes and for 200 mm cubes.

The influence of the loading rate can also be observed, but this effect is not so significant as the size effect. Generally, the rough interface leads to a much better bond between NSC and granite or limestone than the smooth interface, which confirms the findings from the static tests. The highest dynamic splitting tensile strength was found for the rough interface between NSC and granite, while the lowest one was for the smooth interface between NSC and limestone.

Fig. 1 provides a summary of the specific impact energies, obtained by dividing the impact energy by the fracture area (where only the horizontal projection of the area is considered). For limestone the influence of the specimen size on the specific impact energy is very small compared to NSC or granite. This trend can also be seen when the interface bond is tested dynamically: For the limestone-NSC interface there is nearly no size effect, whereas for the granite-NSC interface the effect can be observed. All test results show that the impact energy for the rough granite-NSC interface is higher than that for the rough limestone-NSC interface. Due to the scatter of the results it is difficult to draw any reliable conclusion for the impact energies of the smooth interface connections.

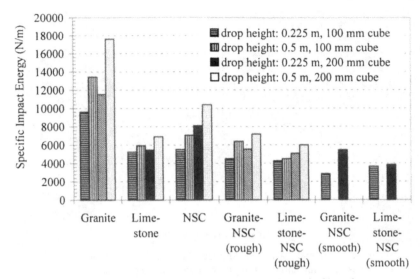

Fig. 1. The specific impact energies of granite, limestone, NSC and of interfaces.

Some examples of the fracture surfaces of the rough interfaces between NSC and rock can be seen in Figs. 2 to 5. All of these photographs show the concrete side of the bimaterial interface. The dark grey areas in Figs. 2 and 3 represent parts of the granite, which remained attached to the concrete after the impact fracture. The light grey areas in Figs. 4 and 5 represent parts of the limestone.

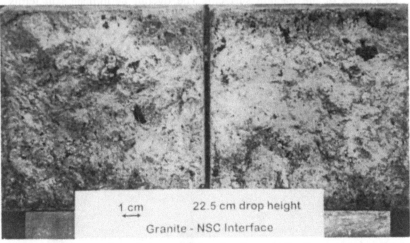

Fig. 2. Two concrete specimen halves tested under the same loading conditions. The dark grey areas are pieces of granite still bonding to the concrete after impact.

Fig. 3. Two concrete specimen halves tested under the same loading conditions. The dark grey areas are pieces of granite still bonding to the concrete after impact.

The following conclusions may be drawn from these photographs: First, an increase in the impact velocity results in a "stronger" bond between the two materials with a rough interface. The crack propagates in a straighter line and does not deflect so much. Therefore the tendency increases for the crack not to follow the interface, but to go right through one of the materials. This effect is even more pronounced in the case of limestone, because of its lower tensile strength, which also can be seen in Fig. 5 (large light grey areas of the fracture surface). The overall performance in terms of the specific impact energy and dynamic splitting tensile strength is better for the granite than for the limestone. This can be explained by the low fracture energy of limestone itself, which is about 50% of that for granite.

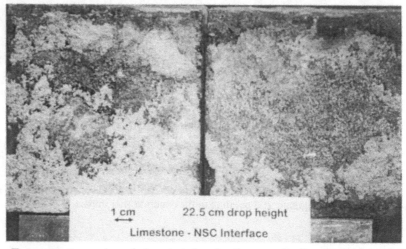

Fig. 4.   Two concrete specimen halves tested under the same loading conditions. The light grey areas are pieces of limestone still bonding to the concrete after impact.

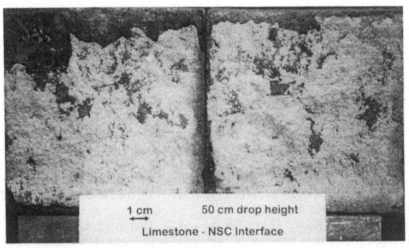

Fig. 5. Two concrete specimen halves tested under the same loading conditions. The light grey areas are pieces of limestone still bonding to the concrete after impact.

4 Conclusions

The following conclusions can be drawn from these test results:
- Fracture mechanics parameters for dynamic loading conditions can be obtained using the splitting tension test configuration. However the specimen size has to be sufficiently large to ensure that these parameters are size independent.
- Impact data for granite, limestone and NSC indicate that these materials are stress-rate sensitive. This is also found for the bonding between NSC and rock. In general, the bonding is stronger under impact loading as compared to static loading.
- The dynamic fracture energy and the splitting tensile strength both strongly depend on the roughness of the bimaterial interface: An increasing roughness leads to higher toughness and strength.
- The bonding of the NSC - granite interface appears to be better, in terms of fracture energy and splitting tensile strength, than the bonding between NSC and limestone.

Acknowledgement

The authors are grateful to the staff of the University of British Columbia, Department of Civil Engineering, Vancouver, Canada where this work was carried out. One of the authors (Dr. Rieder) received a "Schrödinger fellowship" from the Austrian National Science foundation FWF, which he gratefully acknowledges.

References

1. Rice, J.R. Elastic fracture concepts for interfacial cracks. *J. of Appl. Mech. Trans. ASME*, Vol. 55, No.1, 98-103

2. Suo, Z. (1989) *Interface fracture mechanics*. PhD thesis, Harvard Univ., Cambridge, Mass.

3. Hutchinson, J.W. (1990) *Mixed mode fracture mechanics of interfaces*. (ed. M. Ruhle, A.G. Evans, M.F. Ashby, and J.P. Hirth), Pergamon Press, New York, pp. 295-306

4. Stankowski, T., Runesson, K., and Sture, S. (1993) Fracture and slip of interfaces in cementitious composites. I: Characteristics. *J. of Eng. Mech.*, Vol. 119, No. 2, pp. 292-314

5. Stankowski, T., Runesson, K., and Sture, S. (1993) Fracture and slip of interfaces in cementitious composites. I: Implementation. *J. of Eng. Mech.*, Vol. 119, No. 2, pp. 315-27

6. Tschegg, E.K., Kroyer, G., Tan, D.-T., Stanzl-Tschegg, S.E., and Litzka, J. (1995) Investigation of bonding between asphalt layers on road construction. *J. of Trans. Eng.*, Vol. 121, No. 4, pp. 309-16

7. Tschegg, E.K., Rotter, H.M., Roelfstra, P.E., Bourgund, U., Jussel, P. (1995) Fracture mechanical behavior of aggregate-cement matrix interfaces. *J. of Mat. Civil Eng. (ASCE)*, Vol. 7, No. 4, pp. 199

8. Hassanzadeh, M. (1995) Fracture mechanical properties of rocks and mortar/rock interfaces, in *Microstructure of Cement-Based Systems/Bonding and Interfaces in Cementitious Materials* (ed. Diamond, S., Mindess, S., Glasser, F.P., Roberts, L.W., Skalny, J.P., and Wakeley, L.D.), Proceeding MRS Fall Meeting, Boston, Nov. 1994, MRS, Pittsburgh, PA, Vol. 370, pp. 377-86

9. Mindess, S., and Alexander, M. (1993) Mechanical phenomena at cement/aggregate interfaces, in *Materials Science of Concrete IV*, (ed. Skalny, J.P.), The American Ceramic Society, Westerville, OH, pp. 263

10. Alexander, M., and Mindess, S. (1995) Use of chevron-notched cylindrical specimens for paste/rock interface experiments. *J. of Cem. Conc. Res.*, Vol. 25, No. 2, pp. 345

11. Banthia, N., Mindess, S., Bentur, A. and Pigeon, M. (1989) Impact Testing of Concrete Using a Drop-weight Impact Machine. *J. of Exp. Mech.*, Vol. 29, No. 2, pp. 63-9

12. Banthia, N. (1987) *Impact resistance of concrete*. PhD thesis, Univ. of British Columbia, Canada.

13. Bentur, A, Mindess, S, Banthia, N. (1986) The behaviour of concrete under impact loading: Experimental procedures and method of analysis. *Materiaux et Constructions*, Vol. 19, No. 113, pp. 371-8

14. Carneiro, F. L. L. B., and Barcellos, A. (1953) Tensile Strength of Concretes. *RILEM Bulletin* (Paris), No. 13, pp. 97-123

15. Raphael, J.M. (1984) Tensile Strength of Concrete. *ACI Journal, Proceedings*, Vol. 81, No. 2, pp. 158-65

16. Gettu, R., Aguado, A., and Oliveira, M.O.F. (1996) Damage in High Strength Concrete due to monotonic and cyclic compression - A study based on Splitting Tensile Strength. *ACI Materials Journal*, Vol. 93, No. 6, pp. 519-23

17. Bazant, Z.P., Kazemi, M.T., Hasegawa, T. and Mazars, J. (1991) Size Effect in Brazilian Split-Cylinder tests: Measurements and Fracture Analysis. *ACI Materials Journal*, Vol. 88, No. 3, pp. 325-32

PART NINE

EFFECT OF ADDITIVES AND ADMIXTURES ON ITZ STRUCTURE AND PROPERTIES OF CONCRETE

32 CHARACTERS OF INTERFACIAL TRANSITION ZONE IN CEMENT PASTE WITH ADMIXTURES

K. KOBAYASHI, A. HATTORI and T. MIYAGAWA
Department of Civil Engineering, Kyoto University, Kyoto, Japan

Abstract
The interfacial transition zone around aggregate in cement paste with mineral admixtures are investigated with the Vickers hardness test, and the effects of mineral admixtures are discussed in this study.
Keywords: ground granulated blast furnace slag, interfacial transition zone, limestone powder, Vickers hardness

1 Introduction

The areas around aggregate in cement paste, which are looser and greater in pore numbers and sizes, are called "transition zone" or "interfacial transition zone (ITZ)".
It is assumed that the ITZ has such a structure as described below [1]: plates of calcium hydroxide, $Ca(OH)_2$, are crystallized like "card house", and the spaces between them are not completely filled with hydrates, thus making larger and greater volume of capillary pores than the bulk region. It is regarded that the existence of the ITZ around aggregate is the cause of the weaker strength of mortar and concrete than cement paste, and that aggressive agents, which influence durability of these cementitious materials, penetrate toward the inside of these materials through the porous ITZ.

Therefore, it is very important to clarify the structures of the ITZ in order to understand the characters of hardened concrete.

In this study, the ITZ around aggregate in cement paste with the mineral admixtures, ground granulated blast furnace slag (ggbs) and limestone powder (LP) which has not been used frequently as admixtures for concrete, is investigated with the Vickers hardness test. The influences of these admixtures on the structures of the ITZ are discussed.

The Interfacial Transition Zone in Cementitious Composites, edited by A. Katz, A. Bentur,
M. Alexander and G. Arliguie. Published in 1998 by E & FN Spon, 11 New Fetter Lane,
London EC4P 4EE, UK, ISBN: 0 419 24310 0

2 Materials and mixtures

In this paper, the whole amount of cement and mineral admixtures is described as "powder". The physical and chemical characteristics of these powders are presened in Table 1 and 2. The cement was ordinary portland cement, OP. Three types of ggbs and three types of limestone powder were used as admixtures. The effects of specific surface areas of these admixtures on the structure of the ITZ were investigated. Parenthesized numbers after the names of the powders denote their respective specific surface areas.

Table 3 shows the mix proportions of the cement paste in this study. All these mixtures had the same water-powder ratio except for OP6. Mixture OP10 was a standard mixture which contains only OP as powder without admixtures. The name of each mixture normally signs the ratio of OP to admixtures by weight except for mixture OP10 and mixture OP6. In the case of OP6LP4, for example, the powder which consists of OP and LP in the ratio of 6:4 was used. In other words, 40% of OP in the mixture OP10 was replaced by LP. In this study, 20%, 40% and 60% of OP by weight were replaced by LP or ggbs. Mixture OP6 was also the one without admixtures which had a different water-powder ratio from that of the other mixtures, but had the same water-cement (or water-binder) ratio as mixture OP6LP4 in order to investigate the influences of presence of limestone powder. LP(7000) was mainly used in this study. LP(6000) and LP(3000) were used for the mixture OP6LP4, in order to clarify how the specific surface areas or particle diameters of LP affect the characters of the ITZ. The ggbs used mainly in these mixtures was ggbs(8000). Ggbs(6000) and ggbs(4000) were used also in mixture OP6ggbs4. Therefore, 12 types of mixtures were investigated in total.

Table 1. Properties of cement and mineral admixtures

	Specific gravity	Specific surface area (cm^2/g)
OP	3.16	3320
LP(7000)	2.73	6770
LP(6000)	2.73	5840
LP(3000)	2.73	3010
ggbs(8000)	2.90	7950
ggbs(6000)	2.90	5960
ggbs(4000)	2.90	4100

Table 3. Mix proportion of cement paste

	W/P[*](%)	W/B[**](%)	OP:LP:ggbs
OP10	34	34	100:0:0
OP8LP2	34	43	80:20:0
OP6LP4	34	57	60:40:0
OP4LP6	34	69	40:60:0
OP8ggbs2	34	34	80:0:20
OP6ggbs4	34	34	60:0:40
OP4ggbs6	34	34	40:0:60
OP6	57	57	100:0:0

* P(powder)=C+LP+ggbs
**B(binder)=C+ggbs

Table 2. Chemical compositions of cement and mineral admixtures

	Ig.loss	Insol.	SiO_2	Al_2O_3	Fe_2O_3	CaO	MgO	SO_2	Na_2O	K_2O
OP	1.2	0.1	21.7	5.4	3.0	64.1	1.7	1.8	0.29	0.51
LP(7000)	42.5	0.0	0.7	0.0	0.1	54.1	1.4	-	-	-
LP(6000)	42.7	0.0	0.7	0.0	0.1	54.3	1.1	-	-	-
LP(3000)	42.7	0.0	0.6	0.0	0.1	54.2	1.3	-	-	-
ggbs(8000)	0.1	0.2	33.2	14.2	0.2	42.1	6.8	-	0.24	0.31
ggbs(6000)	0.2	0.2	33.5	13.8	0.2	42.1	6.8	-	0.24	0.37
ggbs(4000)	0.1	0.1	33.1	13.9	0.2	42.2	6.7	-	0.20	0.31

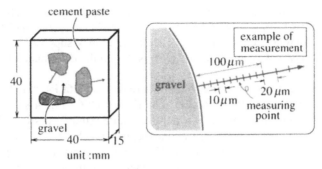

Fig. 1. Vickers hardness measurement

Model specimens consisting only of cement paste and crushed hard sandstone, with the diameter ranging from 10 mm to 15 mm, were used in this study, as it is expected that the adjacent ITZ formed around fine aggregate would combine with each other and then the measurement results could be affected.

3 Experiments

Each cement paste was mixed in a mixer according to JIS R 5201. Afterwards, aggregate was added and mixed again. Every specimen was cast into a 40 mm x 40 mm x 160 mm prism, and wrapped immediately by a polychlorinated vinylidene film to prevent water evaporation. The specimens were demolded after 24 hours from mixing and then cured in water at 20℃ until specified age.

At this age, specimens were cut with an oil cutter into a dimension of 40 mm x 40 mm x 10 mm and then immersed into ethanol to prevent further hydration. The cut surface including aggregate section was polished on a turntable with emery paper of #800~#2400.

Vickers hardness with 2.94×10^{-2} N (3 gf) was used to measure the ITZ around coarse aggregate on the polished surface. Measurements were made every 10 μm up to 100 μm from the interface between the aggregate and the cement paste, and every 20 μm from 100 μm to 200 μm on the line normal to the interface as shown in Fig. 1.

4 Results and discussions

4.1 Effects of limestone powder

Fig. 2 shows the Vickers hardness of the ITZ around aggregate in the mixture OP10, OP6 and that in the cement paste with limestone powder. Fig. 3 shows the minimum value of hardness around aggregate in these mixtures. Weak areas of 50~60 μm width existed around aggregate in all mixtures except for OP4LP6. Also there existed the regular bulk region beyond these weak areas. Vickers hardness of the ITZ and the width of this zone were almost constant with age, while the Vickers hardness of the bulk region increased. In contrast, there did not exist weak areas around aggregate in OP4LP6, presumably because the bulk region was porous due to the high water-cement ratio and had a weak structure similar to the ITZ.

It can be recognized that not only the hardness of bulk region but also the width of ITZ in OP6 differed from those of OP10. It is reported that the width of ITZ decreases with the decline of water-cement ratio [2]. Therefore, it can be supposed that the difference in width of the ITZ between OP10 and OP6 is due to the difference of water-cement ratio.

On the other hand, Vickers hardness of bulk region in OP6LP4, with the same water-cement ratio as OP6, was 15~20 higher than that of OP6. Fig. 4 shows the particle diameter distributions of limestone powder used in this study. As shown in this figure, there were large particles over 100 μm, and marks made with the Vickers diamond needle on the cement paste surface with 2.94×10^{-2} N load were almost within 10 μm. Therefore, it may be assumed that Vickers hardness of cement paste might be affected by not only the tightness of hydrate structures but also the existence of admixtures. In addition, it is reported that the strength of cementitious materials with limestone powder is larger than that without limestone powder in spite of the same water-cement ratio [3]. Accordingly, the increase of hardness of the bulk region might have affected the increase of strength of the cementitious materials, although there does not exist a liner correlation between the strength and the hardness of porous materials because of heterogeneity of them in micro scale.

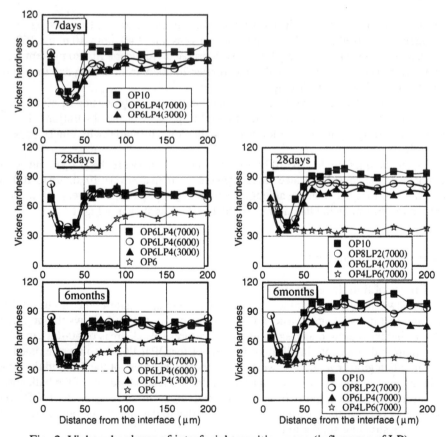

Fig. 2 Vickers hardness of interfacial transition zone (influences of LP)

The width of the ITZ in OP6LP4 was closer to that in OP10, having the same water-powder ratio as OP6LP4, than to that in OP6, having the same water-cement ratio as OP6LP4. Although there are various other theories about the formation mechanism of the ITZ [4], it is suggested that a film of water formed around the aggregate in fresh cement paste obstructs the formation of a dense hydrated structure around the aggregate. So it could be assumed that there existed less free water, which is not bound on powder particle surfaces, because of the high volume of powder in OP6LP4, creating a state similar to the cement paste having a low water-cement ratio [5]. In this study, the width of the ITZ in OP10, OP8LP2 and OP6LP4 was nearly equal to each other and the width had better correlation with the water-powder ratio than with the water-cement ratio as shown in Fig. 5.

The widths of the ITZ in OP6LP4(7000), OP6LP4(6000) and OP6LP4(3000) were very similar to each other, although it could be expected that the amount of free water and the width of the ITZ around aggregate would decrease with the increase of the specific surface area of limestone powder. Therefore, it is impossible to fill the ITZ completely with the addition of limestone powder at least within the range of water-

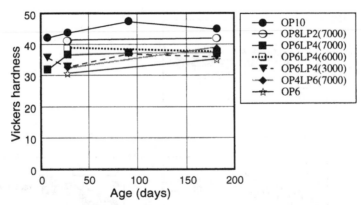

Fig. 3. Minimum value of Vickers hardness of cement paste

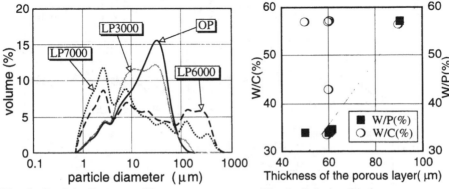

Fig. 4. Particle diameter of limestone powder

Fig. 5. Relationship between thickness of porous layer and W/C or W/P at 28 days

cement ratio examined in this study, although it could be possible to reduce the width of the ITZ to some extent.

It is widely known that the cement hydration is accelerated by limestone powder and that the strength of cementitious materials with limestone powder in early age [for example, 6]. In this study, the Vickers hardness of bulk paste in OP6LP4 and OP4LP6 did not increase considerably after 7 days age. Hence, it was supposed that cement had been hydrated exceedingly before 7 days age and cement hydration did not occur afterwards because of a lack of the unhydrated cement.

4.2 Effects of ground granulated blast furnace slag

Fig. 6 shows Vickers hardness of the ITZ around aggregate in the mixture OP10 and those with ground granulated blast furnace slag. Fig. 7 shows the minimum value of the hardness around aggregate in these mixtures. The hardness of bulk region in the mixtures with ggbs was as high at any ages as those in OP10. Therefore, hydrate was formed due to potential reactivity of ggbs and the bulk region of cement paste with ggbs was as tight as one with OP10 in spite of less volume of cement.

As mentioned above, there existed the ITZ around aggregate in OP10 and the hardness

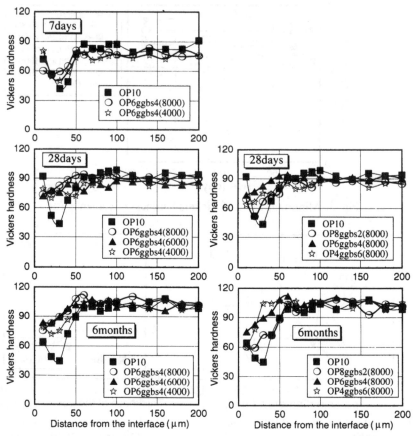

Fig. 6. Vickers hardness of interfacial transition zone (influences of ggbs)

and the width of these weak areas did not change at all after 7 days age. In contrast, the hardness of the ITZ in the mixtures with ggbs were higher than that in OP10. It is widely known that ggbs consume $Ca(OH)_2$ through hydration and fill the large pores in ITZ with hydrates [7]. Thus, it could be considered that the large pores around aggregates were filled with ggbs hydrates in this study.

Although the hardness of the ITZ in OP8ggbs2 was lower than those in the other mixtures with ggbs and was similar to that in OP10 at 28 days age, it increased till 6 months of age or later, with minimum value of 60, in contrast with that in OP10 which did not change at all with age. In addition, the ITZ in OP6ggbs4 or OP4ggbs6 was already slightly harder than that in OP10 at 7 days age and continued to increase their hardness and to decrease their width with age till at least 6 months age. It could also be considered that this hydration reaction already began before 7 days age and continued for a long period afterwards if estimated from the differences in the hardness of the ITZ between OP10 and the mixtures with ggbs.

The hardness of the areas adjacent to the aggregates in the mixtures without ggbs, regardless of use of limestone powder, was very high because of the existence of $Ca(OH)_2$ crystal layer or lateral confinement by aggregate [8]. On the contrary, the hardness of ITZ in the mixtures with ggbs decreased toward the aggregate. It could be assumed that $Ca(OH)_2$ crystals on the aggregate surface were consumed through hydration of ggbs. In addition, there existed gap owing to autogenious shrinkage of cement paste with ggbs between aggregate and cement paste. So lateral confinement by aggregate to cement paste mentioned above might not be efficient.

In this study, the ITZ in mixtures with ggbs with large specific surface areas were harder and narrower than those with small ones at 7 or 28 days age, when they had the same replacement ratio of ggbs, although the hardness of their bulk regions were very similar to each other. In addition, the hardness of the area adjacent to aggregate in OP6ggbs4(4000) was high at 7 days age. In contrast, the hardness of this area in OP6ggbs4(8000) had already become low at 7 days age. Therefore, it could be concluded that the structures of the ITZ could be affected by the specific surface area of ggbs particularly in early ages, and ggbs with larger specific surface area could hydrate early and actively.

In this study, it could be recognized that the porous ITZ was filled with the use of

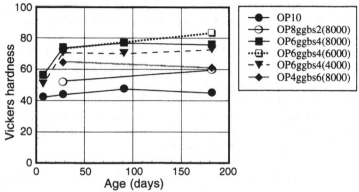

Fig. 7. Minimum value of Vickers hardness of cement paste

ggbs, and dosage and the specific surface area of ggbs affected its density. Therefore, it is necessary to apply an adequate replacement ratio of ggbs to the ordinary portland cement taking water-cement ratio into consideration in order to increase concrete durability.

5 Conclusions

The conclusions obtained in this study are summarized as follows :
1. The width of the interfacial transition zone in cement paste with limestone powder was narrower than that in cement paste without limestone powder, for the same water-cement ratio.
2. The width of the interfacial transition zone in cement paste with limestone powder had a closer relationship to the water-powder ratio than that to the water-cement ratio.
3. Ggbs filled the interfacial transition zone tightly with the hydrate, and this effect was observed from early age.
4. The hardness of the interfacial transition zone enhanced with the increase of replacement ratio of ordinary portland cement with ggbs and of the specific surface area of ggbs.

6 References

1.   Monteiro, P. J. M. (1986) Improvement of the aggregate-cement paste transition zone by grain refinement of hydration products. *8th Int. Cong. on the Chem. of Cement,* Vol. 3. pp. 433-437.
2.   Hanahara, S. (1992) Relationship between pore structure and character of hardened concrete. *Ph.D. thesis,* Keio University, Japan
3.   Sakai, E. (1992) Filler cement (In Japanese) *Cement & Concrete,* No. 546. pp. 129-136.
4.   Breton, D., Carles-Gibergues, A., Ballivy, G. and Grandet, G. (1993) Contribution to the formation mechanism of the transition zone between rock-cement paste. *Cement and Concrete Research,* Vol. 23. pp. 335-346.
5.   Kobayashi, K., Kume, T., Miyagawa, T. and Fujii, M. (1994) Vickers hardness and bond strength of super workable concrete (in Japanese). *2nd Symposium on Self-compacting Concrete.* pp. 149-156.
6.   Kobayashi, K., Hattori, A., Miyagawa, T., and Fujii, M. (1997) Effect of Limestone Powder as Additive on Hydration of Cement in Early Age. *1997 International Conference on Engineering Materials,* Vol. 1. pp. 811-820.
7.   Uchikawa, H., Hanehara, S. and Sawaki, D. (1993) Estimation of the thickness of transition zone in hardened mortar and concrete and investigation of relationship between their thickness and strength development (In Japanese). *Concrete Research and Technology,* Vol. 4, No. 2. pp.1-8.
8.   Igarashi, S. (1993) Structure of interfacial zone and bond mechanism in fiber reinforced cementitious composite materials. *Ph.D. thesis,* Kyoto University, Japan

33 A LOOK AT THE INNER-SURFACE OF ENTRAPPED AIR BUBBLES IN POLYMER–CEMENT–CONCRETE (PCC)

M. PUTERMAN
National Building Research Institute, Technion, Haifa, Israel

Abstract

Polymer-cement-concrete (PCC) is made by the addition of a polymer admixture, in the form of an aqueous dispersion to a cementitious mix. The dispersion is usually stabilized with surfactants and colloidal stabilizers. These surface active agents cause the formation of a large number of air bubbles in the fresh and hardened mix.

Electron microscopy studies have shown that the inner surfaces of these cavities are covered by a dense growth of calcium-hydroxide crystals in the form of well shaped fine platelets which stick out perpendicular to the surface with the crystalline a-axis in the main dimension. The question is raised, and is unanswered, regarding the origin of the growth of these platelets-crystals.

Keywords: air-bubbles, polymer-cement-concrete (PCC), Portlandite.

1 Introduction

Polymers are widely used, nowadays, as admixtures for the modification of various properties of cementitious materials like common mortar and concrete [1-4]. Workability of the fresh paste and ductility and toughness of the hardened mix are among the properties improved by using the polymeric admixtures.

The polymers are usually added to the fresh cementitious mix in the form of an aqueous dispersion (sometimes refereed to as "latex") to a content value around 10-20% w/w in regard to the cement content ($p/c \sim 0.1$-0.2). The polymer dispersion contains also a significant amount of surfactants and colloidal stabilizers intent to stabilize the originally formed emulsion at the polymerization process as well as the resulting dispersion. When mixed into the mortar or concrete mix these surface active

The Interfacial Transition Zone in Cementitious Composites, edited by A. Katz, A. Bentur, M. Alexander and G. Arliguie. Published in 1998 by E & FN Spon, 11 New Fetter Lane, London EC4P 4EE, UK, ISBN: 0 419 24310 0

agents introduce into the mix a large number of closed air bubbles which remain entrapped also in the hardened material. Unless treated by an anti-foam agent the content of the air bubbles might reach quite a high value, up to 30-40% by volume, and more. Most of the bubbles range from a few microns to a few hundreds of microns in size, as can be seen in fig 1.

The formation of a transition zone between cement paste and aggregate grains, in which an enhanced growth of calcium-hydroxide (CH) crystals takes place is a well known phenomenon in concrete microstructure [5-7]. The origin of these CH crystals is from the water solution, saturated with calcium and hydroxyl ions, which fills the pores and cavities around the aggregates. As the hydration process progresses more and more of the these ions - a "product" of the hydration process - are formed and enter the solution. The calcium-hydroxide being only sparingly soluble in water precipitate out of the saturated solution on the available surfaces of the aggregates, as well as on those of the hydrating cement particles. The crystallites grow on the aggregate surface usually as platelets with their main dimension perpendicular to the surface.

The hardness and strength of the calcium- hydroxide ("Portlandite") is much lower than that of the aggregate or the hard cement paste and hence it has an important effect on the mechanical properties of the hardened concrete.

This short communicate intends to report some findings related to the formation and structure of CH crystallites on the inner surface of air bubbles entraped in a polymer-cement-concrete composite. It is part of a larger study concerning properties and structure of PCC.

2 Sample description

The mortar-samples examined and described in this communicate have the following composition:

cement : sand	= 1 : 3	
water : cement	= 0.7	
polymer : cement	= 0.2	

The sand used was fine quartz sand, the polymer was a pure acrylic and the cement - ordinary Portland cement.

3 Calcium-hydroxide crystallites growth

Electron microscopy studies show, as mentioned already, that the PCC hardened phase enclaves a large number of air bubbles ranging from a few microns to a few hundred of microns in size (fig 1). "Looking" into the air bubble reveals that the inner surface of the cavity (...is there at all an "external surface" in this case?) is covered with a dense growth of the Portlandite crystals (fig 2-4). The hexagonal crystals are well shaped as thin (2-3 microns), but rather wide, platelets sticking inwards - into the cavity - to a distance of up to a few tens of microns. It is clear to observe that the platelet-shaped crystals are formed with their a-axis in the platelet main dimension (and the c-axis in the thickness dimension), and are stacked perpendicular to the surface on which they grow.

Fig. 1 Micrograph showing the air bubbles in PCC.

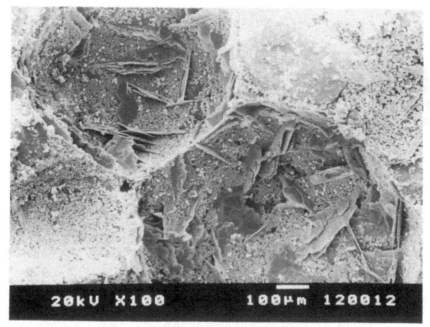

Fig. 2 Micrograph showing two adjacent cavities of air bubbles.

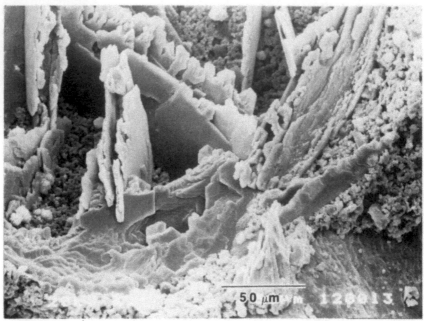

Fig. 3   Micrograph showing part of an inner surface of a cavity covered with growing crystals.

Fig. 4   A close look at crystallites developed on an inner surface of a cavity.

The described findings raise the question regarding the origin and reason of the growth of the large CH crystals. Comparing the growth and shape of these crystallites to those developing on the surface of solid aggregates in concrete mixtures, one can conclude that the two phenomena are very similar. In both cases the crystals grow on the available surface; in the case of the air bubbles the available surface is the "inner" one. The main reason for the growth of the CH crystals on the aggregate is the availability of water saturated with calcium-hydroxide adjacent to its surface. As it seems from the well shaped large crystals, they have grown without interruption in a very favorable environment. This would have required the presence of a saturated solution from which the crystals can grow. One possibility is that during the hydration and hardening process the "empty" bubbles are filled with water saturated with the calcium-hydroxide formed as a hydration product. The fine platelets can then crystallize out of the saturated solution as they do on solid aggregates.

It is believed that these CH crystals, in contrast to the CH crystals in the aggregate-cement paste transition-zone, do not have a particular effect on the mechanical properties of the PCC material as they do not bridge or connect two different phases.

4 References

1. Ohama, Y., Polymer-modified mortars and concretes, Ch. 9, in Concrete Admixtures Handbook, ed. by V. S. Ramachandran, 2nd ed., Noyes Pub., 1995.
2. Polymer Modified Hydraulic Cement Mixtures, ed. by L. A. Kuhlmann and D.G. Walters, ASTM Pub. STP 1176, 1993.
3. Zhao Su, Microstructure of Polymer Cement Concrete, (Doctoral Thesis), Delft University Press, 1995.
4. Handbook of Polymer Modified Concrete and Mortars, Properties and process Technology, by Y. Ohama, Noyes Pub., 1995.
5. Hadley, D.W., The Nature of the Paste-aggregate Interface, Ph.D.Thesis, Purdue University, 1972.
6. Zimbelman, R., A contribution to the problem of cement -aggregate bond, Cem. Concr. Res., vol. 15, 1985. p 801.
7. Monterio, P.M., Maso, J. C. and Olliver, J. P., The aggregate- mortar interface, Cem. Concr. Res., vol. 15, 1985. p. 953.

34 THE ROLE OF SILICA FUME IN MORTARS WITH TWO TYPES OF AGGREGATES

A. GOLDMAN
National Building Research Institute, Technion City, Haifa, Israel
M.D. COHEN
School of Civil Engineering, Purdue University, W. Lafayette, IN, USA

Abstract

The process of transition zone modification in silica fume mortars was studied in this work. The experimental program was developed to distinguish that process from the modification of bulk paste in the same mortars. Gap graded mortars with varying aggregate size, and with constant aggregate volume content, were made. A change of aggregate size could induce a change of the transition zone volume content in mortar. Thus, it was possible to quantify the modification of the transition zone alone, and to investigate separately the effects of silica fume on the transition zone and on the bulk paste.

Mortars with a natural quartzite and with a dolomitic limestone, as well as corresponding pastes, were prepared. Dynamic modulus of elasticity E was measured by a resonant frequency method. The results showed a substantial decrease of E along with an aggregate size increase in reference mortars (without silica fume). This relationship indicated the weakening effect of the transition zone on mortars without silica fume. On the contrary, mortars with silica fume proved very low, if any, sensitivity to changes of aggregate size, which could be related to the modification of the transition zone. In pastes, dynamic modulus of elasticity E did not increase in the presence of silica fume.

Keywords: Mortar, paste, silica fume, transition zone, bulk paste.

1 Introduction

Silica fume can provide an extensive strength increase in portland cement concrete. Several studies concluded that such increase was due to mainly the modification (densification) of the transition zone between bulk paste an aggregate. Microstructure and properties related to the transition zone and its role in the behavior of cementitious

The Interfacial Transition Zone in Cementitious Composites, edited by A. Katz, A. Bentur, M. Alexander and G. Arliguie. Published in 1998 by E & FN Spon, 11 New Fetter Lane, London EC4P 4EE, UK, ISBN: 0 419 24310 0

systems were investigated in numerous works. However, as indicated other studies, it could be more likely that silica fume improved mainly the microstructure and strength of the bulk paste. According to these works, the transition zone densification could not serve as the major process leading to the concrete strengthening.

Regarding the role of the transition zone, it should be mentioned that this issue was addressed in early works [1]. Transition zone - in terms of weak link, or poor bond - between paste and aggregate was investigated years before the effects of silica fume were formulated [2]. The key role of better bond between paste and aggregate in improving the concrete performance was addressed to in numerous studies. In this work, transition zone is referred to as a result of interaction between paste and aggregate in a fresh mix, that is, a zone of paste in a close aggregate vicinity where the paste is not as dense as in the bulk. Total volume of paste in mortar or concrete is simply a sum of bulk paste and transition zone. The initial so called "width" of the transition zone may vary in the range of 5 to 50 μm [3,4].

Investigations related to the role of silica fume included, as a rule, direct measurements of mechanical properties, such as compressive strength. However, other techniques can be applied as well, based on non-destructive measurements. The method developed in this work used measurements of resonant frequency, which is sensitive to density and correlates with the dynamic modulus of elasticity. Partially, the results were reported when this work was in progress [5].

2 Experimental

Two groups of mortars were prepared, with two types of aggregate. Mortars of group A contained a siliceous aggregate, that was a quartzite from Baraboo, Wisconsin. Mortars of group B were made with a dolomitic limestone from Delphi, Indiana. Each group included two series of mortars, with an without silica fume, and correspondent pastes for reference. The specimens were 1"x1"x12" (~ 25x25x300 mm) bars cured in water at ~20°C.

Volumetric approach implemented in this work suggests that volume occupied by the transition zone depends on the aggregate surface area. If a series of mortars is designed with the same mix proportions by mass and by volume, then a mortar with a larger aggregate surface area should contain a larger volume of the transition zone. This is illustrated in Fig. 1. For an equal paste volume content in mortars A and B, total aggregate surface area in mortar B is larger. Accordingly, the volume of the transition zone in this mortar is larger, and the volume of bulk paste is smaller, than in mortar A.

The variety of aggregate surface area was achieved by sieving and separating fine aggregates into size fractions. For each size, mass surface area was calculated as described in a previous publication [5]. The results are shown in Table 1.

The mortars contained: (a) ASTM Type I portland cement; (b) commercial grade silica fume with more than 90% amorphous silica and 24.300 m²/kg. specific surface area; (c) de-ionized water. Water-to-cementitious materials ratio was 0.518 in mortars with quartzite aggregate and 0.457 in mortars with dolomitic limestone aggregate.

The mortars were prepared without water-reducing admixtures. Mix proportions of mortars are given in Table 1.

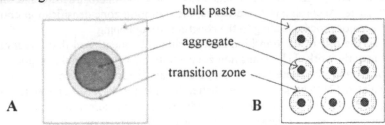

Fig. 1. An illustration of mortars with the same types and amounts of ingredients, but with different aggregate size fractions:
mortar A - larger aggregate size fraction, but lower mass surface area;
mortar B - smaller aggregate size fraction, but higher mass surface area.

The following assumptions were made, as shown in Fig. 1: (a) the aggregate particle shape is spherical, to simplify the calculations, and (b) the average "width" of the transition zone is not influenced by the aggregate size. Some experimental evidence supporting the latter was published [6].

Table 1. Mix proportions of mortars

Mix No.	Aggregates: Type	Size Fraction (mm)	Surface Area (m²/kg)	Content (g)	Cement (g)	Silica Fume (g)	Water (g)
1	Quartzite	4.75 - 2.36	0.6	813	540	60	310.8
2	"	1.18 - 0.60	2.5	813	540	60	310.8
3	"	0.30 - 0.15	10.1	813	540	60	310.8
4	"	4.75 - 2.36	0.6	800	600	-	310.8
5	"	2.36 - 1.18	1.3	800	600	-	310.8
6	"	1.18 - 0.60	2.5	800	600	-	310.8
7	"	0.60 - 0.30	5.0	800	600	-	310.8
8	"	0.30 - 0.15	10.1	800	600	-	310.8
9	Limestone	4.75 - 2.36	0.6	813	540	60	274.0
10	"	1.18 - 0.60	2.5	813	540	60	274.0
11	"	0.30 - 0.15	10.1	813	540	60	274.0
12	"	4.75 - 2.36	0.6	800	600	-	274.0
13	"	2.36 - 1.18	1.3	800	600	-	274.0
14	"	1.18 - 0.60	2.5	800	600	-	274.0
15	"	0.60 - 0.30	5.0	800	600	-	274.0
16	"	0.30 - 0.15	10.1	800	600	-	274.0

Paste specimens were prepared, corresponding to each series of mortars. In two paste mixes (with and without silica fume), water-to-cementitious materials ratio was 0.518, same as in mortars with quartzite aggregate, while in two other pastes it was 0.457, identical to mortars with dolomitic limestone. Mix proportions, size, and curing conditions of paste specimens were the same as of mortar specimens. Two specimens of each mortar and paste mix were made. Dynamic modulus of elasticity E was determined by the resonant frequency technique at the ages of 1, 3, 7, 14, 28, and 56 days for all specimens, except for a 14 days testing of mortars with dolomitic limestone that was omitted.

3 Results

In all reference dolomitic limestone aggregate mortars (without silica fume), the results showed a clear tendency of E to drop substantially, along with the increase of the aggregate surface area (Fig. 2A). At all ages, a somewhat steeper drop of E in these mortars took place when the aggregate surface area was lower, between 0.6 and 1.3 m^2/kg. A moderate drop could be associated with a relatively high aggregate surface area, 5.0 to 10.0 m^2/kg (Fig. 2A).

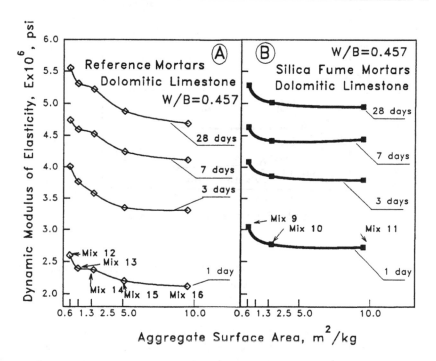

Fig. 2. Correlation between dynamic modulus of elasticity E and aggregate surface area in dolomitic limestone mortars: A- reference mortars; B- silica fume mortars.

Some reduction of E in silica fume mortars took place only when the aggregate surface area was 0.6 to 1.3 m²/kg (Fig 2b), as it was observed in reference mortars, but to a lower extent. In contrast with reference mortars, there was actually no drop of E for all other surface areas. This was repeatedly observed at all ages, starting with 1 day. As one could see, the E values remained the same, despite the aggregate surface area varied between 1.3 m²/kg and 10.0 m²/kg. Silica fume mortars showed no sensitivity to changes of aggregate size in that wide range.

Apparently, silica fume mortars with dolomitic limestone aggregate were denser than correspondent reference mortars (Fig. 3). Substantial difference in density was observed at 1 day. That difference was gradually smaller at later ages, and was closed between 28 and 56 days (see also Fig. 2B).

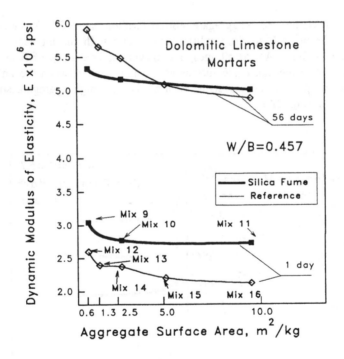

Fig. 3. Correlation between dynamic modulus of elasticity and aggregate surface area in mortars with dolomitic limestone at 1 day and at 56 days.

A very similar phenomenon was observed in mortars with quartzite aggregate, reported previously [5]. Dynamic modulus of elasticity E dropped substantially in reference mortars (Fig. 4). The drop was steeper at lower aggregate surface area, 0.6 to 1.3 m²/kg, but moderate at higher surface area. On the contrary, similar values of E, with actually no reduction, were obtained in silica fume mortars, indicating no sensitivity to changes of aggregate surface area.

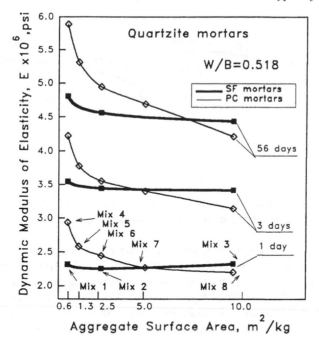

Fig. 4. Correlation between dynamic modulus of elasticity and aggregate surface area in mortars with quartzite aggregate.

Unlike in mortars with dolomitic limestone, silica fume mortars with quartzite appeared not to be denser than correspondent reference mortars (Fig. 4). This might be due to mix design, where silica fume was introduced by replacement of portland cement by weight. Although mortars with dolomitic limestone were prepared in the same way, their density in the presence of silica fume - at 1, 3, and 7 days - was higher than of reference mortars (Figs. 2, 3).

All silica fume mortars made with dolomitic limestone were denser than correspondent mortars with quartzite (compare results in Fig. 3 and in Fig. 4). This difference between two groups of silica fume mortars, as well as the relative increase of density in silica fume mortars with dolomitic limestone (described above), possibly indicates a stronger contribution of dolomitic limestone, which was denser than quartzite. Bulk unit weights of these aggregates were 2.72 t/m^3 and 2.65 t/m^3 respectively.

The results, where average density of silica fume mortars with dolomitic limestone became similar to that of reference mortars, were obtained later on, at 28 and 56 days, as shown in Figs. 2, 3. Although microstructure studies have not been included here, this can probably be related to densification of reference mortars as a result of cement hydration. Then, reference mortars could become denser than silica fume mortars at later ages.

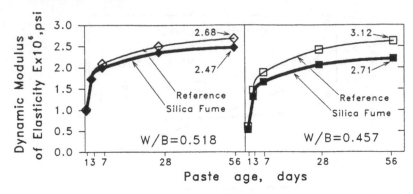

Fig. 5. Development of dynamic nodulus of elasticity in two groups of pastes, prepared with water-to-cementitious materials ratio W/B=0.518, and W/B=0.457.

In two pastes made along with a group of mortars containing dolomitic limestone, where water-to-cementitious materials ratio W/B was 0.457, dynamic modulus of elasticity E was higher than in two other pastes, corresponding to quartzite mortars, W/B=0.518 (Fig. 5), which indicated higher density. This was in accordance with the difference in W/B, that was higher in quartzite mortars. In both groups, values of E in reference pastes were consistently higher than in silica fume pastes, also indicating higher density.

4 Summary

Silica fume mortars with dolomitic limestone aggregate exhibited very low, if any, sensitivity to variation of aggregate surface area in a wide range. Same result was obtained for quartzite aggregate. On the other hand, all reference mortars demonstrated high sensitivity in the same range of aggregate surface area. Silica fume pastes, that represented bulk paste, were not denser than reference pastes. The drastic reduction of sensitivity was therefore a result of paste densification in the transition zone due to the presence of silica fume. The densification took place already at 1 day, which should be related to the microfiller effect of silica fume, formulated in previous works [6,7].

A small, although consistent, decrease of dynamic modulus of elasticity E was observed at low aggregate surface area corresponding to the size fraction of 4.75 to 2.36 mm. This result could be taken as an experimental error, as was suggested previously [5]. On the other hand, consistency of this observation, combined with a steep drop of E in the same range for reference mortars, might indicate the influence of the aggregate size. Consequently, the transition zone modification process might be more efficient for smaller aggregate size fractions.

In the presence of silica fume, use of a denser aggregate - dolomitic limestone - led to a denser mortar. This illustrated the active role of aggregate due to the transition zone modification, expressed this time in terms of density.

5 Acknowledgments

The authors are thankful to National Science Foundation and Indiana Department of Transportation for partial support of this research. The helpful comments of Professor Shin-ichi Igarashi are appreciated.

6 References

1. Farran, J. (1996) Introduction: The Transition Zone - Discovery and Development, *Interfacial Transition Zone in Concrete*, RILEM Report 11, Toulouse, pp. xiii-xv.
2. Glucklich, J. (1972) The Strength of Concrete as a Composite Material, *Proceedings of International Conference on Composite Materials*, Japan, Vol. 4, pp. 104-112.
3. Goldman, A. and Bentur, A. (1989) Bond Effects in High-Strength Silica-Fume Concretes, *ACI Materials Journal*, Vol. 86, No. 5, pp. 440-447.
4. Bentur, A. Goldman, A. and Cohen, M. D. (1987) The Contribution of the Transition Zone to the Strength of High Quality Silica Fume Concretes, *MRS Proceedings*, Vol. 114, Boston, pp. 97-103.
5. Cohen, M. D., Goldman, A. and Wai-Fah Chen (1994) The Role of Silica Fume in Mortar: Transition Zone Versus Bulk Paste Modification, *Cement and Concrete Research*, Vol. 24, pp. 95-98.
6. Goldman, A. and Bentur, A. (1992) Effects of Pozzolanic and Non-Reactive Microfillers on the Transition Zone in High Strength Concretes, *Interfaces in Cementitious Composites*, RILEM Proceedings 8, Toulouse, pp. 53-61.
7. Goldman, A. and Bentur, A. (1993) The Influence of Microfillers on Enhancement of Concrete Strength, *Cement and Concrete Research*, Vol. 23, pp. 962-972.

Acknowledgments

The authors are grateful to National Science Foundation and Defense Department of Transportation for partial support of this research. The helpful comments of Professor ... are acknowledged.

References

PART TEN
CLOSING CONCLUSIONS

35 ITZ STRUCTURE AND ITS INFLUENCE ON ENGINEERING AND TRANSPORT PROPERTIES: CONCLUDING REMARKS TO THE CONFERENCE

A. BENTUR
National Building Research Institute, Faculty of Civil Engineering,
Technion, Israel Institute of Technology, Haifa, Israel

1 Introduction

The main object of the conference was to resolve the role of the ITZ in controlling properties of engineering significance, in particular (i) mechanical characteristics of concretes and cementitious composites and (ii) the role of ITZ in durability performance, through its influence on transport properties. An implied assumption prior to the conference was that the nature of the ITZ is well established and defined. However, the debate sparked after the keynote lecture by Sidney Diamond (1) indicated that this is not the case. Thus, the conference was interlaced with continuous discussions that involved two main issues: the nature of the ITZ and its role in controlling engineering properties and durability. The following remarks are intended to present an overview of some of the key questions that were highlighted during these discussions and provide my personal input as to the concepts that may have to be modified or implemented if we are to address some of the apparently conflicting notions and data presented at the conference.

2 ITZ - Reality or Myth?

1. The treatment of ITZ in various publications suggests that it is a well defined materials parameter. Thus the variety of microstructures observed for systems of similar overall compositions has lead to some confusion, and resulted in questioning of the whole concept. However, if we address the ITZ as a systems parameter which is dependent on the overall composition as well the method by which the material was made (i.e. manufacturing process), than many of the apparent inconsistencies can be understood, as highlighted in the next remark.

2. In view of the above, the ITZ microstructure for a given composition should be addressed in terms of a range of structures that might be observed for a given material. An illustration of this concept is the variety of microstructures that can be obtained

The Interfacial Transition Zone in Cementitious Composites, edited by A. Katz, A. Bentur,
M. Alexander and G. Arliguie. Published in 1998 by E & FN Spon, 11 New Fetter Lane,
London EC4P 4EE, UK, ISBN: 0 419 24310 0

between two extremes: a composite specimen in which paste is cast against the plane surface of a rock and a concrete made from similar ingredients which were well mixed. In the previous case (composite specimen) bleeding and wall effects may be crucial in controlling the interfacial microstructure, while in the latter the two may be eliminated or significantly reduced by intensive mixing and gradation of aggregates, resulting in a different ITZ microstructure. This concept may be true for other systems such as fiber reinforced cements which may be produced by different manufacturing methods. Thus, the ITZ should not be assigned with a definite and distinct microstructure (2).

3. An additional modification of the above concepts may be based on a view of the paste microstructure from a different perspective, which is that many microstructural characteristics observed are due to inhomogeneity. Thus, the ITZ microstructure may be addressed as inhomogeneity near the interface, which is superimposed on the bulk matrix inhomogeneity. Perhaps, the aggregate is only "magnifying" inhomogeneity due to the "wall effect", but is not generating anything which is fundamentally different. If so, than the term Interfacial Transition Zone (ITZ) should be reconsidered and addressed as Interfacial Inhomogeneity.

4. The changes in concepts and apparent contradictions pointed above, should be viewed from the perspective of the evolution of the ITZ concepts. In the past, due to limitations of instrumentation for micro-analysis of composition and microstructure, the interfacial zone had to be studied by semi-destructive methods using composite specimens (e.g. Toulouse method). With the advent of instrumentation for in-situ micro-analysis, the actual composite could be characterized, and distinct differences could be observed between the ITZ of composite specimens and the ITZ in the actual material. However, many of the concepts are still based on the initial observations made in composite specimens.

3 Parameters affecting ITZ Microstructure and its Properties

1. In view of the above and the ability for in-situ microanalysis of the ITZ, we should re-examine some of the basic notions of the factors controlling ITZ microstructure. Parameters which will generally reduce the inhomogeneity of the bulk paste are expected to have a similar, and perhaps even greater effect on the ITZ, if the latter is viewed as interfacial inhomogeneity (see 3section 2). Indeed, additions of pozzolans, silica fume, water reducing admixtures and intensive mixing are known to reduce the size of the ITZ and reduce the difference between its microstructure and that of the bulk paste.

2. In view of above, one may realize why it is sometimes difficult to resolve the practical influences of the ITZ. Characteristics of ITZ evaluated in composite specimens (interfacial bond, diffusion at interfacial zones) are not necessarily representative of the actual material where the ITZ may be different; indirect evaluation if the influences of ITZ by testing of the actual concrete may be difficult to interpret, as parameters which influence the ITZ also influence the bulk matrix (see 1 in this section).

4 Influences of ITZ on Engineering and Transport Properties

1. In particulate composites (e.g. concrete) effects of ITZ on engineering properties can amount to an increase in strength and modulus of elasticity in the range of 20 to 40%, mainly by mobilizing the aggregates to become reinforcing inclusions. From a practical point of view this is important, although not drastic.

2. Modeling of the influence of ITZ on transport properties, assuming a model of aggregates with a rim of ITZ which is more porous than the bulk, suggest that there is a threshold aggregate content, above which the transport should increase drastically. Experimental studies (e.g. reference 3) do not provide a clear cut support for this prediction. Additional factors might mitigate this effect, amongst others being dilution of the paste content with impervious aggregates as well as tortuosity influences induced by the presence of aggregates.

On the other hand, the ITZ may have indirect effect on transport properties by creating a zone which can be more readily damaged by microcracking (e.g. reference 4). These influences of ITZ have not been thoroughly evaluated, as rarely transport properties are studied on concretes which have been exposed to influences which may simulate damage induced in field conditions.

3. The ITZ is expected to have considerable influence on the performance of fiber reinforced cementitious composites. Evaluation of this influence by testing bond in pull-out specimens may be misleading as these are essentially composite specimens where the ITZ microstructure is different than that in the actual composite. This is perhaps of no practical significance in macro-fibers (0.1mm diameter and more) where the bond is quite small (due to the small surface area) so that in practice anchoring is achieved not by interfacial effects but rather by mechanical anchoring induced by the crimped shape of commercial fibers. However, in microfibers (where the diameter is of the same order of magnitude as that of the cement grains) the ITZ microstructure is extremely important as it may control the overall mode of failure of the composite, being either ductile or brittle. Changes in the ITZ microstructure in such system with time may induce a change in the mode of fracture over time from ductile to brittle, causing long term performance problems.

5 References

1. Diamond, S. and Huang, J., (1998), *The Interfacial Transition Zone: Reality or Myth*, Proceedings of the 2nd International Conference on "The Interfacial Transition Zone in Cementitious Composites", A.Katz, A.Bentur, M.Alexander and G.Arliguie (editors), (This book).
2. Bonen, D., (1998), *Features of the Interfacial Transition Zone and its Role in Secondary Mineralization, pp. 203-212*, ibid.
3. Carcasses, E., Petit, J.Y. and Ollivier, J.P., (1998), *Gas Permeability of Mortars in Relation with the Microstructure of Interfacial Transition Zone (ITZ)*, ibid.
4. Arliguie, G., Francois, R. and Konin, A., (1998), *Modeling the Modification of Chloride Diffusion in Relation to the ITZ Damage*, ibid.

Author Index

Abipramono R. 67
Alexander M.G. 292
Andrade C. 187
Arliguie G. 93, 196

Ballivy G. 114
Banthia N. 216
Bartos J.M. 234
Bažantová Z. 276
Belaid F. 196
Bellanger M. 125
Bentur A. 152, 267, 335
Bentz D.P. 43
Besari M.S. 67
Bijen J.M.J.M. 59
Boiko S.V. 75
Bonen D. 224

Cabrillac R. 163
Campbell K. 216
Carcasses M. 85
Chen S.-W. 243
Cohen M.D. 324
Courard L. 207

Darimont A. 207
De Rooij M.R. 59
Delagrave A. 103
Diamond S. 3, 141

Fournier B. 114
François R. 93,196
Frens G. 59

Gallias J.L. 163, 171
Garboczi E.J. 43
Goldman A. 324

Hattori A. 311
Homand F. 125
Huang J. 3

Israel D. 251

Kawamura M. 179
Kobayashi K. 311
Konin A. 93
Kovler K. 152

Lee C.H. 259
Lee R.J. 141
Lee S. 259
Legrand C. 283
Li X.S. 259
Li Z.J. 259

Marchand J. 103
Menéndez E. 133, 187
Mindess S. 301
Miyagawa T. 311
Modrý S. 276

Odler I. 152
Ollivier J.P. 85

Perbix W. 251
Petit J.Y. 85
Pigeon M. 103
Puterman M. 319

Qi M. 259

Renaud-Casbonne F. 125
Rieder K.-A. 301
Rivard P. 114

Sanjuan M.A. 187
Schneider U. 243
Scrivener K. 292
Shtakelberg D.I. 75
Singhal D. 179
Skalny J.P. 141
Suarjana M. 67

Trtik P. 234

Tsuji Y. 179

van Mier J.G.M. 51

Vervuurt A. 51

Wasserman R. 267

Wirquin E. 283

Yao W. 259

Key Word Index

Active thin section 59
Adherence 207
Adhesion 207
Aggregates 267, 292
Aging 152
Air bubbles 319
Alkali–silica
  gel layer 179
  reaction 114, 133, 179
Appetency 207

Bond
  properties 259
  strength 276
BSE microscopy 133, 187
Bulk phase 324
Bundled reinforcement 234

Calcareous aggregates 125
Calcium chloroferrite hydrate 163
Calcium ferrite hydrate 163, 171
Calcium hydroxide 179
Calcium leaching 103
Cement 152, 207
  microstructure 224
  paste 59, 163, 171, 267
  –rock composites 276
Ceramic tile 259
Chloride 93, 187
Chloride ions 103, 163
Coagulational contact 75
Concrete 43, 133, 207, 301
  damage 114
  durability 141
  model 67
  strength 267
Corroded reinforcement 163, 171
Corrosion 196
  layer 163, 171
  of reinforcement 179
Cover 187

Crack propagation 234
Crushing strength 67
Cylindrical steel inclusion 67

Damage 93
Deposition site 224
Destruction 75
Deterioration 141
Diffusion coefficient 93
Diffusivity 43
Disruption 75
Durability 125, 152, 196

Expanded shale 243

Fiber reinforcement 216
Fiber-reinforced composites 152
Fly ash 267
Formation 59
Fracture 301
Freezing and thawing cycles 125
Frost resistance 125

Galvanized steel 196
Gas permeability 85
Glass 152
Granite 133
GRC 234
Grooved surfaces 67
Ground granulated blast furnace slag
  311
Gypsum-free cement 276

Hardening 75
High performance concrete 251
Hydration 283

Image analysis 114
Impact 301
In-situ three-point bending test 234
Interaction 75

Interface 141, 243, 301
Interfacial
  bond 216
  inhomogeneity 336
  zones 67
Interfacial transition zone (ITZ) 43, 85,
    93, 103, 114, 163, 171, 179, 196,
    224, 276, 283, 292, 311, 336
ITZ *see* Interfacial transition zone

Kinetics 207

Lightweight 267
  aggregate 243
  aggregate concrete 243
Limestone and siliceous aggregates 283
Limestone powder 311

Materials parameter 335
Mechanical behavior of concrete 243
Metal aggregates 67
Microcement 251
Microcracking 75, 114
Microhardness 243, 276
Microstrength test 43, 59, , 141, 163,
    171, 187, 234, 251, 292
Model 85, 251
Model specimen 67
Modelling 43
Mortar 187, 324
Multi-scale 43

Non-linear stress analysis 67

Particle size 251
Paste 324

Paste–aggregate contact 125
Pellicular moisture 75
Plastic shrinkage cracking 216
Polymer–cement–concrete (PCC) 319
Portlandite 319
Potsdam sandstone 114
Pressure 259
Push off test 259

Reaction rim 114
Reinforcement 187
Repair 207
Rock 301

Sand volume fraction 85
Scanning electron microscope 276
Sea water 179
Secondary mineralization 224
SEM microscopy 133, 152
Shotcrete 216
Silica fume 103, 259, 324
Solid solution 163
Stiffness 292
Strength 283
Succion 207
Sulfate attack 141
Superplasticizer 283
Syneresis 59
Systems parameter 335

Tensile strength 93
Transition zone 243, 324

Vickers hardness 311

Water/binder ratio 103

This index is compiled from the keywords provided by the authors of these papers, edited and extended as appropriate. The numbers refer to the first page of the relevant papers.